A Handbook of Statistical Graphics

A Handbook of Statistical Graphics

Dibyojyoti Bhattacharjee
Senior Lecturer
Department of Statistics
Gauhati University, Guwahati

Kishore K. Das
Reader
Department of Statistics
Gauhati University, Guwahati

Publishing for One World
NEW AGE INTERNATIONAL (P) LIMITED, PUBLISHERS
(formerly Wiley Eastern Ltd.)
New Delhi · Bangalore · Chennai · Cochin · Guwahati · Hyderabad
Jalandhar · Kolkata · Lucknow · Mumbai · Ranchi
Visit us at **www.newagepublishers.com**

Published by New Age International (P) Ltd., Publishers
First Edition: 2008

ISBN (13) : 978-81-224-1227-9

ISBN (10) : 81-224-2127-X

Rs. 100.00

C-07-08-1977

Printed in India at Nisha Enterprises, Delhi.
Typeset through Author.

PUBLISHING FOR ONE WORLD

NEW AGE INTERNATIONAL (P) LIMITED, PUBLISHERS
(formerly Wiley Eastern Limited)
4835/24, Ansari Road, Daryaganj, New Delhi - 110002
Visit us at **www.newagepublishers.com**

In Memory of

LATE JATINDRA CHANDRA DAS

Preface

The title "A Handbook of Statistical Graphics", itself indicates the very purpose of the book. Data visualization or graphical representation of Statistical data is now an emerging subfield of Statistics. Today, graphical methods play an important role in all aspects of statistical investigation, it begins with explanatory plots, supports various stages of analysis and helps in the final communication and display of results. These days we see extensive use of graphics in print media, television news, sports coverage, advertisement etc., and this gives us an idea about the development that is taking place in the field of statistical graphics. Most of the recent versions of statistical software are now producing high-resolution self-explanatory graphics and is adding number of more graphs in every recent version of their product. However the softwares do not provide sufficient text to support the purpose and interpretation of the relatively less common graphics and so their use is restricted. Now one can get ones research data analyzed by using graphical tools that are handy and easier compared to other statistical tools.

But absence of proper text has hindered the development of graphics and restricted their use even though the software is available for producing the graphs. Especially in India we could not quote a single title which is dedicated to Statistical Graphics. The graphical techniques if used can reduce a lot of calculations that is generally involved with other statistical techniques in reaching to a conclusion.

Various universities of Europe and USA have started to develop specialized courses on "Data Visualization" or "Statistical Graphics". The trend is soon going to enter different parts of the globe and this book can be considered a handy material for "Statistical Graphics".

The book comprises of various statistical plots that can be used for analysis, display and comparison of data. Data sets are provided with most of the plots and relevant calculation, if any, are also shown. The variables considered along the axes are highlighted and the interpretation of the graph is also discussed. The uses of each of the graphical tool are also put forward. At the initial stage a brief history about the development of graphical techniques, supported with an extended bibliography is provided. For fresher, some commonly used terms in statistical graphics are also made available.

Dibyojyoti Bhattacharjee

Kishore K. Das

Contents

Chapter 1

The Genesis

1.1 INTRODUCTION

Today statistical charts and graphs, play an important role in data presentation and analysis. They are used in our textbooks and classrooms, they are playing an increasing role in government reports, they appear daily in our newspapers and popular magazines, in television news and sports coverage. With the advancement in data visualization (a graphical tool used for understanding abstract relationships among variables), statistical graphics were developed and used to interpret the meaning of data collected in many disciplines like biology, data mining, digital imaging, educational tests, finance and marketing, survival analysis, engineering, medicine, health care and so on. Such graphical techniques are also helpful in behavioural sciences, particularly when the data is studied under several categories. Thus it is essential to know the ups and downs that statistical graphics has undergone over the years and reach its present stage of zenith.

The history of statistical graphics has a deep-rooted link with the history of statistics. The statistical thinking compliments the development in statistical graphics and viceversa. Till the middle of the twentieth century the importance of graphics was connected with the line of statistical thinking, but it was the development of the technology, which largely dictated the progress of statistical graphics and data visualization. Today, quantitative graphics are re-emerging as an important statistical tool. To learn about the tradition and history of the development of statistical graphics one should go through the works of Beniger and Robyn (1978), an excellent study, which comes out with extensive bibliography. Friendly and Dennis (2002) performed a very exhaustive work - with lots of photographic displays of remarkable historic graphs, their creators, consisting as many as 228 references.

Here for studying the history of statistical graphics, a century wise study is made starting from seventeenth century. There is a big history of the subject prior to that, a brief note of that has been added, just to provide earlier context against which the later developments can be viewed.

1.2 PRIOR TO 1600

The graphs and diagrams that we see around us today had their origin in geometric diagrams; some of them were probably drawn to understand the positions of stars and other celestial bodies, and in the making of maps to help travellers in navigation. In the 16th century, techniques and instruments for precise observation and measurement of physical quantities were well-developed. The idea of representing various features with the aid of mathematical functions gained momentum by that time, which triggered the beginning of the husbandry of visualization.

There are a number of claimants for the earliest known diagrammatic representation. Beniger and Robyn (1978) states that, the earliest known map is present on a clay tablet dated at 3800 B.C. which depicts all of Northern Mesopotamia with conventions and symbols still used. But according to Friendly and Dennis (2002), the oldest known map dates back to 6200 B.C., which is the "*Map of Konya Town*", available at the Museum at Konya, Turkey.

Even during the 3200 B.C., Egyptian surveyors used to study their land in terms of coordinates not quiet different from the Cartesian system that is in use. By the 1500 B.C. in Egypt a lot of practical problems were solved by the application of geometry, including graphical representation.

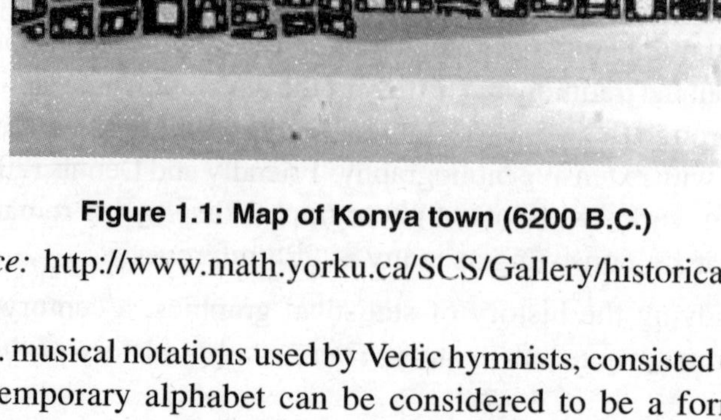

Figure 1.1: Map of Konya town (6200 B.C.)

Source: http://www.math.yorku.ca/SCS/Gallery/historical.html

In 700 B.C. musical notations used by Vedic hymnists, consisted of 300 symbols, based on contemporary alphabet can be considered to be a form of graphical representation, quiet in keeping with the Franconian reforms by the 13th century A.D. Euclid the famous Greek astronomer and mathematician by 300 B.C. contributed in the field through formal statements of geometric principles. By the tenth century A.D., medieval astronomers expressed planetary movements as cyclic lines on spatial temporal grids, the diagrams thus formed are similar to the modern line graphs.

A French mathematician Nicole Oresme (1482) used Proto Bar graph developed in 1350 to study the logical relation between tabulating values. The use of the rectangular coordinates probably, for the first time finds its expression in the works of the great Leanardo da Vinci (1500) to analyse the velocity of falling objects.

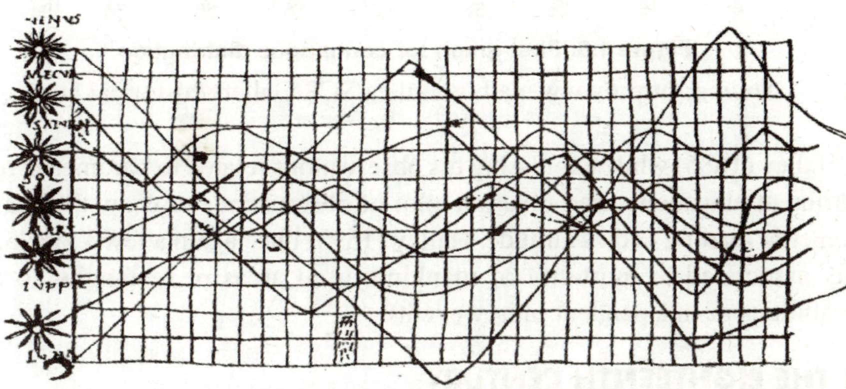

Figure 1.2: Planetary movements drawn in the tenth century A.D.

1.3 SEVENTEENTH CENTURY

The 17th century saw great new growth in theory and their applications, the rise of analytic geometry, theories of errors of measurement, the birth of probability theory, and the beginning of demographic statistics and their application in real life and even in fields like politics. The coordinate system was revisited in mathematics, analytic geometry, relationship established between graphed line and equation by Descartes, Fermat and other Franch mathematicians in the first half of 17th century. At the end of that century, the necessary elements were at hand- some real data of significant interest, some theory to make sense of them, and a few ideas for their visual representation.

The first major success using coordinates belonged to Halley (1686), who plotted barometric reading against elevation above sea label, then fitting a hyperbolic curve to the scatterplot. First visual representation of statistical data was made by Langren (1644) while studying variations in determination of longitude between Toledo and Rome.

Christian Huygens in 1669 drew the first graph of a continuous distribution function, a graph for life table and demonstrated how to find the median remaining lifetime for a person of given age.

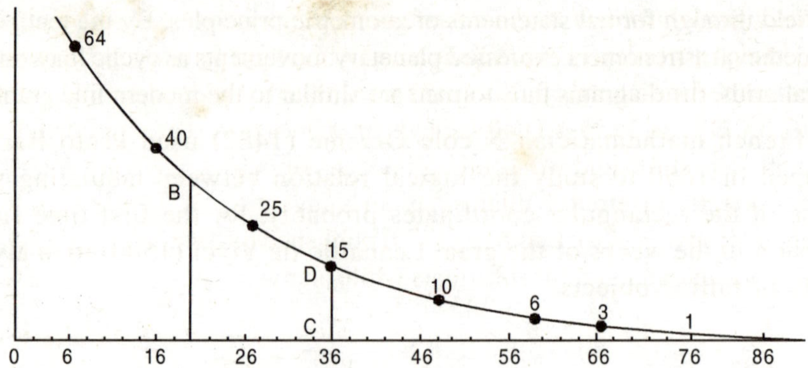

Figure 1.3: First graph for continuous distribution.

Source: http://www.math.yorku.ca/SCS/Gallery/historical.html

Halley (1686) while expressing his observations related to barometric readings at different elevations used bivariate plot of observations to show the changes in barometric reading versus altitude. Halley (1693) once again a few years latter in a study associated to mortality of mankind, used areas of rectangles to display probabilities of independent binary events.

1.4 THE EIGHTEENTH CENTURY

By the beginning of the eighteenth century lot of numerical information were collected from politics and economy, which needed to be represented in a way such that the data speaks to the eyes. Most of the well-known forms of graphs and techniques that are used today were established during this period. For example, bivariate plots of data points, used since the middle of 17th century, line graphs of time series data since 1724, curve fitting and interpolation (1760), the notion of measurement error from as deviation from a regular graphed line (1765), graphical analysis of periodic variation (1779), statistical mapping (1782), bar charts (1786) and printed coordinate paper (1794). Actually in that century, what the researchers felt was that, the growth of the seeds of visualization that were lodged previously. Advances in the field of map-making was also noticed in that century- as new graphic forms like isolines and contours were invented. Also the technological innovations made during that period-like colored printing, lithography and other developments eased the task of creating such visual representations.

At the beginning of the 18th century, Halley (1701) made a remarkable contribution to cartography by developing the isobar map, lines of equal magnetic declination for the world. Hauksbee (1712) while performing some experiments on hydro dynamics saw some interesting figures like, hyperbola on the ascent of water between two glass planes and developed the Literal line graph. Cruquius (1724) developed the abstract line graph. Lambert (1760) performed curve-fitting and interpolated points from empirical data points. Bayes (1763) in his famous "*Essay towards solving a problem in the doctrine of chance"* drew the graph of the beta density. Joseph Priestley (1765) made a early breakthrough when he published a historical time line-chart of life spans of 2,000 renowned people who lived between1200 B.C. to 1750 A.D. Priestley's time lines proved a commercial success and popular sensation, and went through dozens of editions, the last in 1820. The time lines directly inspired William Playfair's invention of the bar chart in 1786 (Beniger and Robyn, 1978). Lambert (1779) published the first semi-graphic display combining tabular and graphical formats while analysing periodic variation in soil temperature.

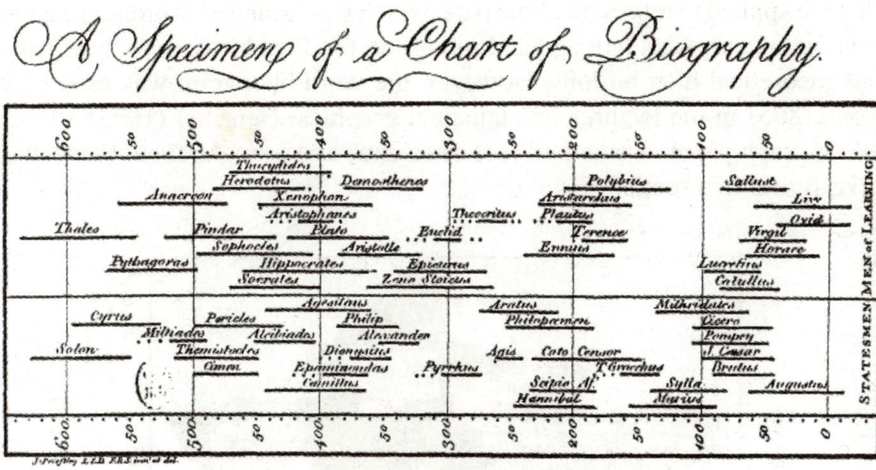

Figure 1.4: Priestley's time line chart

*Source:*http://www.math.yorku.ca/SCS/Gallery/historical.html

1.5 NINETEENTH CENTURY

The use of Priestley's time-line graphs kept on increasing in the subsequent editions of Playfair's *Atlas*. Experimenting with that Playfair (1801) invented two new types of charts that can be used for the purpose of comparison, pie chart and circle graph. Feinberg (1979) commented that, though graphics had a long history but it was only after the works of Playfair in the late 18th and the early 19th century that the use of graphs and charts for data display became accepted practice. Soon after that, Alexander Von Humboldt (1811) combined the concept of the bar and pie chart to produce the sub-divided bar graph. In the same paper he also introduced a variation of Playfair's circle graphs, superimposed squares for the comparison of areas.

The graph paper used for plotting was made available commercially for the first time by, Dr Buston in London in 1794. But in published form such papers occurred six years latter in the works of Howard (1800). Lalanne (1845, 1846) later introduced both logarithmic grid and polar coordinates. Jevons (1863) developed the semi-logarithmic paper.

Duplin in 1819 (there is some controversy relating to the date) produced a Cartogram, map with shadings from black to white depending on the distribution and intensity of illiteracy in France was probably, the first modern statistical map. By 1820's, graphs and charts started to get published in various scientific journals in Europe. Fourier (1821) drew a bar chart representing the population of Paris by age groupings, and then placed the bars one atop the other to form the ordinate of a first quadrant graph. The abscissa contained the same age demarcations but at equal intervals. By connecting like age groupings with lines perpendicular to the axes, Fourier completed the first graph of a cumulative frequency distribution, which was later called ogive by Galton in 1875. Then Guerry came up with bar charts which were applied to represent crime data, which was arranged in ordered categories for continuous variables like age and month, to produce histograms suggestive of various theoretical distributions. However, the word histogram was used by Karl Pearson (1895) in his lectures on statistical graphics. Herschel (1833) advocated the use of graph paper as standard tool of science and brought about the technique of curve-fitting to a scatter plot.

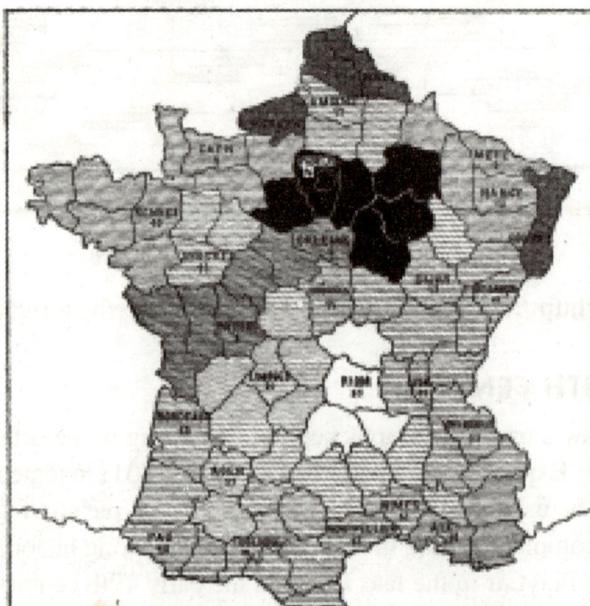

Figure 1.5: First statistical map

Source: http://www.math.yorku.ca/SCS/Gallery/historical.html

A remarkable contribution in this field was made by Quetelet (1827) and by Walker (1874), Superintendent of the U.S. Census. Walker developed age pyramid and bilateral frequency polygon, forms, which he introduced in *Statistical Atlas of the United States*, published in 1874. By the mid of 19th century, application of graphics in statistics became quite commonplace. In 1849, Fletcher published the first statistical map in a statistical journal. Minard used statistical diagrams incorporated in a map with circles proportional to coal production in 1851(published in 1861). In 1857, Florence Nightingale after witnessing deplorable sanitary conditions in the Crimea, wrote *Notes on Matters Affecting the Health, Efficiency and Hospital Administration of the British Army*, which, included several graphs of her own design, which she called "Coxcombs". The figure makes it abundantly clear that far more deaths were attributable to non-battle causes (preventable causes) than to battle-related causes.

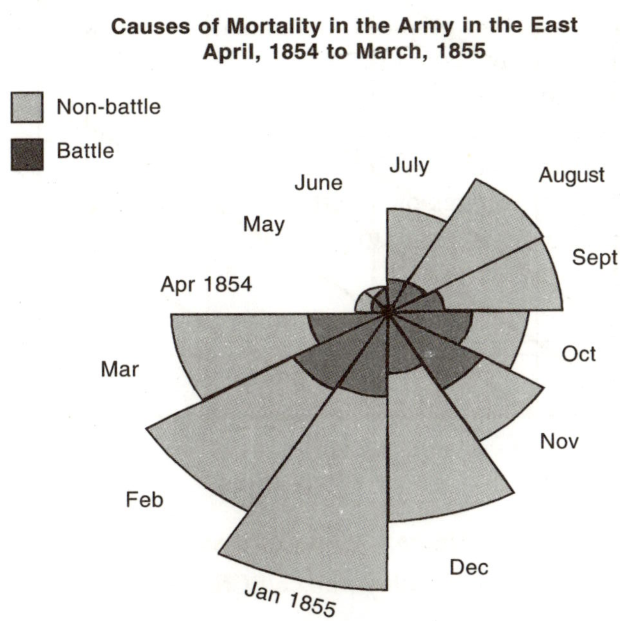

Figure 1.6: The Coxcomb chart of Florence Nightingale a replica from Friendly and Dennis (2002)

Source: http://www.math.yorku.ca/SCS/Gallery/historical.html

The Third International Statistical Congress in Vienna in 1857, organized an exhibition display of graphs and cartograms and explained their merits over other techniques of representing data. Galton (1861) developed the modern weather map, a chart showing area of similar air pressure and barometric changes by means of glyphs displayed on a map.

In 1872, in the ninth census report of the USA Government statistical graphics was most extensively used, which took graphics to common people. At the 8th session of the *International Statistical Congress* meetings at St. Petersburg, Schwabe (1872) classified statistical graphical treatments by form, with consideration of appropriate uses of color, graphical elements and limitations of perception.

The first contour map of physical geography, was published by Buache (1752), and such maps became very popular. Later, Lalanne (1845) used the contour form to reconstruct a three-way table, which give mean temperatures by hour of the day throughout the year. Lalanne plotted months and hours on the abscissa and ordinate, then connected points of similar temperature to form contour lines.

The first use of bilateral histogram and bilateral frequency polygon express themselves, in the U.S. National Atlas by Francis A. Walker (1874) Superintendent of the U.S. Census, USA, when he drew the age pyramid.

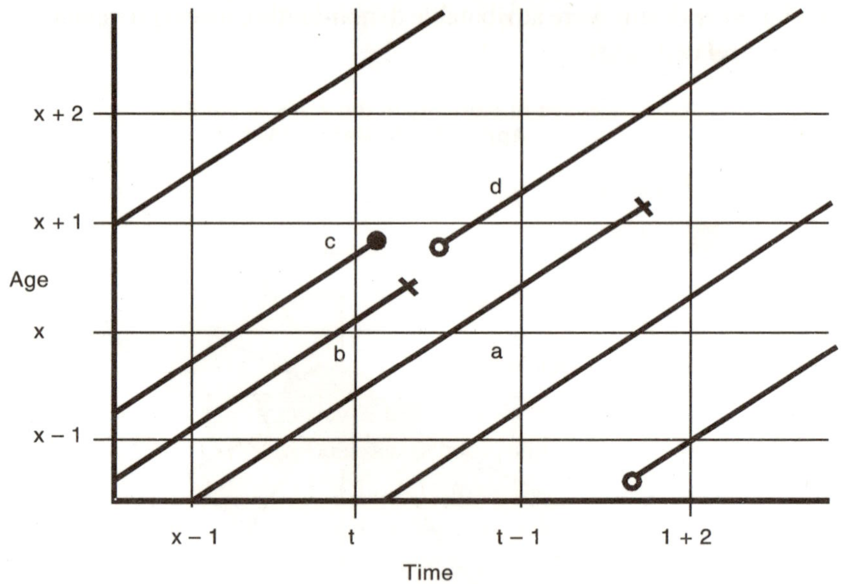

Figure 1.7: A Lexis diagram

Source: http://www.math.yorku.ca/SCS/Gallery/historical.html

Lexis (1875) of Germany put forward a type of diagram named after him as the Lexis diagram, however, some authors questioned the paternity of this diagram. The Lexis diagram expresses the relationship among age, calendar time, and life spans of individuals simultaneously.

Etienne-Jules Marey, 1830-1906, was among the pioneers of dynamic graphics. Dynamic graphics are graphical representations, which are animated, have some movement and possess dynamic phenomena. Marey also used and developed many devices to record and visualize motion and dynamic phenomena: walking, running, jumping, falling, of humans, horses, cats; heart rate, pulse rate, breathing, etc.

By the end of the 19th century, as more statistical data became available, the limitations of two dimensions of the plane for the representation of data were becoming more apparent, especially, relating to interrelationship among at least three variables specially in the field of vital statistics. It led to a graphical problem of representing multivariate distribution and correlations.

Maurice d'Ocagne (1899) of France developed the concept of alignment diagrams, using sets of parallel axes, rather than axes at right angles. Galton (1899)

developed the idea for "log-square" paper, so that normal probability curve appears as a straight line when plotted.

Figure 1.8: The first dynamic graph

Source: http://www.math.yorku.ca/SCS/Gallery/historical.html

1.6 TWENTIETH CENTURY

Though in the nineteenth century we saw great advances in statistical graphics, but in the first half of the twentieth century the progress was hindered, the scene completely changed and graphics took the back seat. However, during 1930s some new innovations were made in the field of statistical graphics. In that period statistical graphics became "main stream". Beniger and Robyn (1978) in a nutshell also describes the history of academic courses and books in statistical graphics produced during that period. At the beginning of the 20th century, statistical graphics had begun to diffuse through textbooks, college courses, mass media into popular domain. One reason was the popularity of the pictogram, which, is a comparative form of a chart based on similar drawings of different sizes. This was developed by Mulhall (1884). Statistical diagrams was first used in school textbooks in France, beginning with Lavasseur's *La France, Avec Ses Colonies* published in 1868. Graphs on mathematical functions began to appear in U.S. textbooks after 1902.

But, with the various advanced methodologies put forward by great statisticians like Fisher, Hotteling, Kendell, etc. many people started to believe that numbers, parameter estimates and standard errors were precise where as pictures were- well, just pictures: pretty or evocative, perhaps, but incapable of stating a 'fact' to three or more decimals. But things started to turn around with John W. Tukey of USA who invented a wide variety of new, simple, and effective graphic displays, under the rubric of "Exploratory Data Analysis". That work was widely circulated around 1970, although not published until 1977, began to make graphical data analysis both interesting and respectable again.

By the end of 1970s significant intersections and collaborations would begin between computer science and statistical graphics. With these technologies becoming prominent, we saw the development software tools like C language, UNIX, etc. at Bell Laboratories and elsewhere, to it was combined the forces and development in data analysis and with the use of modern display and input technology (pen plotters, graphic terminals, digitizer tablets, the mouse, etc.) lead to an explosive growth in new visualization methods and techniques. It is however, a very hard task to provide sufficient overview of the most recent developments in data visualization, for they are so varied and scattered, have occurred very fast, and occurred over various range of disciplines. These developments in visualization methods and techniques possibly depended on advances in the theoretical and technological communications.

Lorenz (1905) published the famous Lorenz curve, which compares the percentiles of cumulative distribution. This probably inspired Wilks and Gananadesikan (1968) to produce the *p-p* plots used for testing the goodness of fit. By that time graphics found extensive use in texts related to physical, chemical, medical and other natural sciences. Moseley (1913) of England, discovered the concept of atomic number which was based largely on graphical analysis (plot of serial numbers of the elements *vs.* square root of frequencies from X-ray spectra).

Statistical graphs related to temperature and population trends appeared in textbooks about 1910; bar and pie charts between 1915-1918. The first textbook in English entirely devoted to graphics was J.B. Peddle's – *The Construction of Graphical Charts,* published in 1910. Soon, Brinton (1914), Gilman (1917), Haskell (1919), Karsten (1925), Palmer (1921) brought about more texts on graphics. The first American college to start a course on graphical methods was the department of agricultural engineering at Iowa State in the academic year 1913-1914, and was taught by M.F.P Castelloe. In 1916, F.J. Warne offered a correspondence course, twenty weekly lessons in fifty dollars, out of the Southern Building in Washington, D.C. Cubberly taught a stanford course, "Graphical Methods for Presenting Facts", annually during the period 1918-1933.

A U.S. government chart book in 1919 entitled, *The War with Germany,* consisted a lot of graphs, which helped in the popularizing of the graphical methods in the United States amongst the masses. Soon, the use of statistical graphics started to get importance through the works of various authors like Eells (1926), von Huhn (1927), Croxton and collaborators (1927,1927,1932), Washburne (1927) etc. in *Journal of American Statistical Association.* The American Society of Mechanical Engineers felt the need of setting up of standards for graphical representation and formed a joint committee of American scientific societies and compiled one such standard, which was published in *JASA* in1915.

Beniger and Robyn (1978) opines that with the advent of World War II, suddenly the importance and growing interest in statistical graphics suffered a hindrance as the academic statisticians became more mathematically conscious. Explaining the reason for this decline in the use of graphics in the middle Fienberg states that with the path breaking works on statistical theory by R.A. Fisher in the 20's and 30's and with the contribution of Hotelling, Neyman, Egon Pearson and others the attention turned towards the rise of mathematical statistics and hence a relative decline to the use of graphs. This trend lasted till mid-1960's, with the developments in computer technology, which made possible the manipulation and representation of large multivariate data sets.

The remarkable graphical approaches that we see about that period consists of Edger Anderson's (1957) circular glyph to represent multivariate data, Bachi (1968),

Andrews (1972) Fourier series to generate multivariate plots, Chernoff's cartoons of a human face to represent multivariate observations, Flienberg's (1975) "Floating Four-Fold Circular Display".

Graphical techniques for displaying of two way contingency tables were developed by Cox and Lauh (1967). Snee (1974) displayed the contingency table graphically which enabled the data analyst to determine which rows or columns are contributing towards the non homogeneity (if the chi square test statistics is significant) Fienberg (1975) used the "Four-Fold Circular Display" to represent 2 × 2 table.

Cleveland and Kleiner (1975) developed a graphical technique for enhancing scatterplots with moving statistics, like the midmean and lower and upper semi-midmean. Ehrenberg (1975) in an unpublished personal communication brought about a comparison: Experimental tests of statistical graphics versus tables. Wainer and Reiser (1976) assessed the effectiveness of various visual displays. They developed the "Cartesian rectangle" to represent 2 × 2 table and experimentally tested it, against other forms of such displays.

Becker and Chambers at the AT&T Bell Labs, USA (1978) developed a language and environment for statistical computation and graphics and called it *S*, which was later sold as a commercial package, *S-Plus*. *S-Plus* later became a *very important tool* for statistical computation and graphics. Snee (1974) made an attempt to represent the frequency of a two way contingency table graphically. Later Hartigan and Kleiner (1981) used Mosaic display for even a multiway contingency table.

Wainer (1974) developed plots like hanging histograms, rootograms, suspended rootograms etc. that can be used for the purpose of empirical validation of normal distribution. Hoaglin (1980) derived a graphical technique, similar to probability plotting, which can be used to judge whether a Poisson model is appropriate for an observed frequency distribution. This plot was called Poissonness plot can be equally applied to truncated Poisson situations.

Figure 1.9: Chernoff faces

Source: http://epce.ed.ac.uk/computing/training/document_archive/

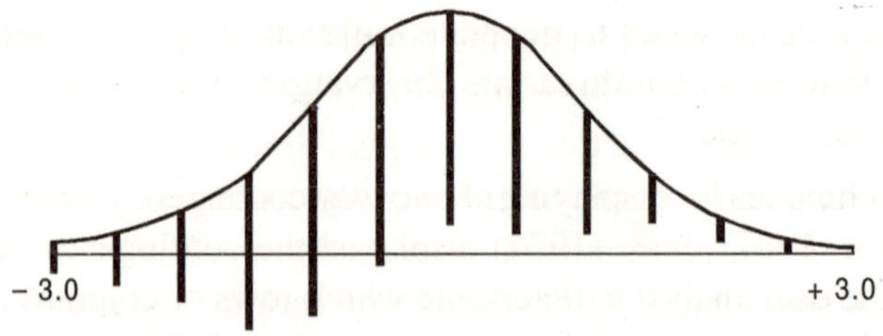

Figure1.10: A hanging histobar from Wainer (1974)

Tukey (1981) came out with a remarkable graphical display of data sets called the 'draftsman display' for three-variables (leading soon to the 'scatterplot matrix') and initial ideas for conditional plots and sectioning (leading later to 'coplots' and 'trellis displays'). Tufte (1983) defined and illustrated some concepts like 'data-ink' ratio, 'lie factor' etc. which was basically designed to study the consistency of graphs and diagrams and plotted out some common errors in it. Inselberg (1985) used the parallel coordinates plots for high-dimensional data.

In the next few years a number of interactive modes for handling large sets of data graphically was developed, of which mention can be made of Becker and Cleveland (1987) who used the concept of brushing, linking and other forms of interaction.

Buja, Asimov, Hurley and McDonald (1988) made the first inclusion of grand tours in an interactive system, which led to visual inference from graphics, interactive scaling of plots, etc. along with linked brushing, linked identification etc. Unwin and Wills (1988) developed interactive graphics for multiple time series with direct manipulation like zoom, rescale, overlaying, etc. Tierney (1990) developed, LISP-STAT an object-oriented environment which supported both statistical computing as well as dynamic graphics. Tukey and Tukey (1990) used textured dot strips to display empirical distributions. Keiding (1990) used lexis diagram in statistical inference and lexis pencil in display of the multivariate data. Johnson and Shneiderman (1991) for studying hierarchical information structures used treemaps. Friendly (1992) with the aid of mosaic display developed a visual analysis tool for log-linear models. Dykes (1996) developed dynamic maps for cartographic visualization, which is basically a map visualization toolkit with graphical tools for viewing data, including a wide range of mapping options. Wilkinson (1999) put forward a comprehensive summary of grammatical rules, which should be followed for drawing graphs.

By the beginning of 1990's statistical graphics and other related tools of visualization gained its popularity. People form engineering and scientific discipline understood that computer based graphics can be used as a problem solving as well as communicative tool. Advanced computing technologies have only increased the importance of graphics as a means of communication, yet technical graphics courses (if any are required) are often the only place students get any formal instruction in developing their ability to effectively use graphics (Clark, Wiebe and Shown, 1996).

The start was probably provided by the North California State University, where a course entitled "*Graphic Communication Program*". The course was started with an aim to explore the future direction of technical graphics in higher education and to learn how graphics can be used to enhance the understanding of technical and scientific concepts. Soon after, several universities in US and Europe started to include courses on statistical graphics as part of various under-graduate and post-graduate courses on science and technology. Such courses were termed as 'Scientific Visualization", "Information Visualization", "Data Visualization" etc. Mention can be made of the following universities who offer related course, University of Utah, Indiana University, University of Massachusetts, Simon Fraser University, The Ohio State University etc. of North America, University of Kaiserslautern, University of Rostock, University of Freiburg, etc. of Germany. Since most of the universities and educational institutes of Europe does not make their course outline available to the public which restricted from further information being collected about other institutes of Europe.

Most of these graphics developed recently are the innovations made by modern computer technology. Graphics drawn with the aid of computers are generally more accurate than hand drawn once. Most of the modern graphical tools have little practical value except when executed by computer. In future, the development of statistical graphics partially or to a great extent will depend on the development of computer graphics hardware and software. The strength of visualization through interactive visualization packages, has enhanced the use of statistics in other fields like medical science, astronomy, engineering sciences etc.

1.7 CONCLUSION

Graphical and diagrammatic representation in no longer a superfluous decoration but it plays an essential role in organizing and visualizing numerical information in statistical terms to a manageable form. Very soon graphics will be more frequently used in analysis of data and inferring from them, instead of being tools of data visualizing only. Newer and newer graphical techniques are designed everyday to improve the quality, accuracy and satisfaction of the statistical data analysis process. With the accessibility to technological advances like color computer graphics, computer animation, dynamic graphs, three-dimensional computer graphics, holography etc. one may see the solutions through innovative use of both graphical and semi-graphical tools. In very recent future, we may come across high-resolution statistical software producing self-explanatory graphics. The extensive use of graphics in print media, television news, sports coverage etc. gives us an idea about its development. Thus from the discussion it can be assumed that data analysis systems of the 21st century will incorporate many advanced graphical methods that will be of help to the data analyst to have a more perceptive, productive and rewarding practice, about the data.

REFERENCES

Anderson, E. (1957). A semigraphical method for the analysis of complex problems. *Proceedings of the National Academy of Sciences, 13* (3): 923–927. Reprinted in *Technometrics*, 2:387–391 (August 1960).

Andrews, D.F. (1972). Plots of high dimensional data. *Biometrics,* 28, 125–136.

Bachi, R. (1968). *Graphical Rational Patterns, A New Approach to Graphical Presentation of Statistics*, Jerusalem: Israel Universities Press.

Beniger, J.R. and Robyn, D.L. (1978). Quantitative graphics in Statistics: A brief history. *The American Statistician*, Vol. 32, 1, 1–11.

Brinton, W. C. (1914). *Graphic Methods for Presenting Facts*. New York: The Engineering Magazine Company. ISBN 0405135041. Xi, 371 p. : ill. ; 24 cm.; reprinted: Arno Press, NY, 1980.

Buache, P. (1752). Essai de géographie physique. *Mémoires de L'Académie Royale des Sciences*, 399–416.

Clark, A.C., Wiebe, E.N. and Shown, T. (1996). Future Directions for Graphics: A Look at the New Technical Graphics Curriculum in NC High Schools. http://www4.ncsu.edu/~wiebe/articles/svc-1996.pdf

Cleveland, W.S. and Kleiner, B. (1975). A graphical technique for enhancing scatter plots with moving Statistics. In *Proceedings of the Annual Meeting*. American Statistical Association, Atlanta, GA.

Cox, D.R. and Lauh, E. (1967). A not on graphical analysis of multidimensional contingency tables. *Technometrics,* 9, 481–488.

Croxton, F.E. (1927). Further studies in the graphic use of circles and bars. *Journal of the American Statistical Association, 22*: 36–39.

Croxton, F.E. and Stein, H. (1932). Graphic comparisons by bars, squares, circles and cubes. *Journal of the American Statistical Association*, *2*: 54–60.

Croxton, F. E. and Stryker, R. E. (1927). Bar charts versus circle diagrams. *Journal of the American Statistical Association*, *22* :473–482.

Cruquius, N. (1724). Observationes accuratae captae ... circa mediam barometri altitudinem, mediam thermometri elevationem, tum et hydrometri varietatem mediam, etc. *Philosophical Transactions*, 4–7.

d'Ocagne, M. (1899). *Traité de nomographie: Théorie des Abaques, Applications Pratiques*. Paris: Gauthier-Villars.

Duplin, C. (1819). Carte de la France éclairée et de la France obscure.

Dykes, J.A. (1996). Dynamic maps for spatial science, a unified approach to cartographic visualization. In D. Parker (ed.), *Innovations in GIS 3*, 177–187. London: Taylor & Francis.

Ehrenberg, A.S.C. (1975). Numerical information processing project. London Graduate School of Business Studies. Unpublished personal communication.

Eells, W.C. (1926). The relative merits of circles and bars for representing component parts. *Journal of the American Statistical Association, 21*, 119–132.

Fienberg, S.E. (1975). *Perspective Canada As a Social Report.* Tech. rep., Department of Applied Statistics, University of Minnesota, Unpublished paper.

Fienberg, S.E. (1979). Graphical Methods in Statistics. *The American Statistician*, Vol. 33, *4*: 165–178.

Friendly, M. (1992). Graphical methods for categorical data. *Proceedings of the SAS User's Group International Conference, 17* :1367–1373.

Friendly, M. and Dennis, D.J. (2002). Milestone in the history of thematic cartography, statistical graphics and data visualization. http://www.math.yorku.ca/SCS/Gallery.

Fourier, J. B. J. (1821). Notions generales, sur la population. *Recherches Statistiques sur la Ville de Paris et le Departement de la Seine, 1*: 1–70.

Galton, F. (1899). A geometric determination of the median value of a system of normal variants from two of its centiles. *Nature, 16*: 102–104.

Gilman, S. (1917). *Graphic Charts for Business Man*, Chicago.

Halley, E. (1686). On the height of the mercury in the barometer at different elevations above the surface of the earth, and on the rising and falling of the mercury on the change of weather. *Philosophical Transactions,* 104–115.

Halley, E. (1693). An estimate of the degrees of mortality of mankind, drawn from curious tables of the births and funerals at the city of Breslaw, with an attempt to ascertain the price of annuities on lives. *Philosophical Transactions, 17:* 596–610.

Halley, E.(1701). The description and uses of a new, and correct sea-chart of the whole world, shewing variations of the compass. London.

Hartigan, J.A. and Kleiner, B. (1981). Mosaics for contingency tables. In W. F. Eddy (ed.), *Computer Science and Statistics: Proceedings of the 13th Symposium on the Interface*, 268–273. New York, NY: Springer-Verlag.

Haskell, A.C. (1919). *How to Make and Use Graphic Charts*. New York: Codex.

van Langren and Florent, M. (1644). La Verdadera Longitud por Mary Toledo.

Hauksbee, F. (1712). Account of the experiment on the ascent of water between two glass planes in an hyperbolick figure. *Philosophical Transactions*, 539–540.

Herschel, J. F. W. (1833). On the investigation of the orbits of revolving double stars. *Memoirs of the Royal Astronomical Society, 5* :171–172. Read January 13, 1832.

Hoaglin, D.C. (1980). A Poissonness plot. *The American Statistician,* Volume 34, 3, 146–149.

Howard, L. (1800). On a Periodical Variation of the Berometer, Apparently due to the influence of the Sun and Moon on the Atmosphere. *Philosophical Magazine, 7*: 355–363.

von Huhn, R. (1927). A discussion of the Eells' experiment. *Journal of the American Statistical Association, 22:* 31–36.

Humboldt, F.W.K.H.A. von (1811). *Essai Politique sur le Royaume de la Nouvelle-Espagne, avec un Atlas Physique et Geographique,* Paris.

Inselberg, A. (1985). The plane with parallel coordinates. *The Visual Computer, 1* :69–91.

Jevons, W.S. (1863). *A Serious Fall in the Value of Gold Ascertained, and Its Social Effects Set Fourth,* London.

Karsten, K.G. (1925). *Charts and Graphs. An Introduction to Graphic Methods in the Control and and Analysis of Statistics.* New York: Prentice Hall.

Lalanne, L. (1845). Sur la Representation Graphique des Tableaux Meteorologiques et des lois Naturelles en General , Appendix to Cours Complet Meteorologie de L.F. Kaemtz, translated and annotated by C. Martins, Paris.

Lalanne, L. (1846). *Memoire sur les Tables Graphiques et sur la Geometrie Anamorphique Appliquees a Diverses Questions qui se Rattachent a l'Art de l'Ingenieur*, Paris.

Lambert, J.H. (1779). Pyrometrie; oder, vom maasse des feuers und der wärme mit acht kupfertafeln. Berlin.

Lexis, W. (1875). Einleitung in der theorie der bevölkerungsstatistik.

Lorenz, M.O. (1905). Methods of measuring the concentration of wealth. *Publications of the American Statisical Association*, 9, 209–219.

Moseley, H. (1913). The high frequency spectra of the elements. *Philosophical Magazine*, pp. 1024-. (continued, 1914, pp. 703-).

Mulhall, M.G. (1884). Dictionary of Statistics. London.

Oresme, N. (1482). Tractatus de latitudinibus formarum. Padova. British Library: IA 3Q024.

Palmer, A.R. (1921). *The Use of Graphs in Commerce and Industry*. London, U.K.

Pearson, K. (1895). Contributions to the Mathemetical Theory of Evolution –II, Skew Variation in Homogeneous Material. *Philosophical Transactions,* 186, Pt. 1, 343–414.

Quetelet, A. (1827). "Recherches sur la Population, les Naissances, etc., dans le Royaume des Pays-Bas," *Nouveaux Memoires de l'Academie Royale Des sciences et Belles-Lettres de Bruxelles, 4,* 117–174.

Schwabe, H. (1872). Theorie der graphischen darstellungen. In P. Séménov (ed.), *Proceedings of the International Statistical Congress*, 8th Session, Pt. 1, 61–73. St. Petersburg: Trenké & Fusnot.

Snee, R.D. (1974). Graphical display of two-way contingency tables. *The American Statistician*, 28, 9–12.

Tufte, E.R. (1983). *The Visual Display of Quantitative Information* . Cheshire, CT: Graphics Press.

Tukey, P.A. and Tukey, J. W. (1981). Graphical display of data sets in three or more dimensions. In V. Barnett (ed.), *Interpreting Multivariate Data.* Chichester, U.K.: Wiley and Sons.

da Vinci, L. (1500). *Notebooks*, Vol. M, Verso 40. Paris: Manuscripts of the Institute of France. (c. 1500).

Wainer, H. (1974). The suspended rootogram and other visual displays: An empirical validation. *The American Statistician,* Vol. 28, No. 4, 143–145.

Wainer, H. and Reiser, M. (1976). Assessing the efficacy of visual displays. In *Proceedings of the Social Statistics Section,* Vol. 19, 89–92, American Statistical Association.

Walker, F. A. (1874). *Statistical Atlas of the United States, Based on the Results of Ninth Census, 1870, with Contributions from Many Eminent Men of Science and Several Departments of the [Federal] Government.* New York: Julius Bien.

Washburne, J.N. (1927). An experimental study of various graphic, tabular and textual methods of presenting quantitative material. *Journal of Educational Psychology, 18* :361–376, 465–476.

Wilkinson, L. (1999). *The Grammar of Graphics.* New York: Springer. ISBN 0-387-98774-6.

Wilk, M.B., and Gananadesikan, R. (1968). Probability plotting methods of analysis of data. *Biometrika,* 55, 1–17.

Chapter 2

Some Terms and Terminology

2.1 INTRODUCTION

In order to be an expert graphic designer it is important to grasp the principles and practice of chart designing starting from the ordinary one-dimensional bar diagram to the complicacies of multi-dimensional graphics. However, the basis of graphs is the rectilinear coordinate system to which most of the users are accustomed to. But diagrams on the other hand do not require the use of such a coordinate system. Let us first concentrate on some of the features that are essential for the display of a chart.

2.2 CHARACTERISTICS OF A GOOD GRAPH

A proper graph should have certain characteristics. They are:

(i) **Accuracy:** it is difficult to produce high degree of accuracy of numerical values in a chart; still an attempt should be made to reach the highest degree of accuracy that can be achieved. Optical illusions and distortions should be avoided as far as practicable. Use of three dimensional bars, circles etc. should be avoided. As the representation is mostly made in a two dimensional plane (be it a piece of paper or computer screen) so a three-dimensional paper mostly contributes to visual illusion. An inaccurate or distorted graph acts against the reputation of representation of data via charts.

(ii) **Simplicity and Clarity:** An important attribute of a chart should be its simplicity. Uses of complicated symbols, unnecessary attachment of superfluous decoration should be minimized and completely avoided if possible. An acceptable chart should not look like a puzzle rather it must be such that the viewer gets the grasp of the information contained in it within few moments on visualizing of the chart. Since charts have become commonplace in the newspapers and television shows so it should be easily readable and in that way must reach the reader more easily. Such displays also help in increasing the interest of the readers as the trend or/ and patterns become clear from the charts. The term clarity here means that the message that is forwarded by a display should be meaningful and comprehensive. It is always recommended not to try for displaying too much in a single graph. Many graphic designers when exposed to a huge data set are tempted to do so and accordingly fail to maintain the clarity of the display.

(iii) **Appearance with Accessories:** The appearance of a chart is the most important step as it is the actual feature that makes or mars the good faith of the graph. A graph has a number of features and a set of accessories associated with it which makes the graph more explanatory. Some of the accessories like legends, title, axes labels etc. (to be explained in the following section) help the graphic designer to develop a better communication with the reader.

2.3 TERMS RELATED TO RECTILINEAR COORDINATE SYSTEM

Of the different type of charts it is the graphs based on rectilinear coordinate system that are extensively used. In order to provide the reader an insight into the various features of such charts the following terminologies may be looked into.

(i) **Coordinate Axes:** The basic structure of the rectilinear coordinate axes consists of two axes the *X*-axis or the abscissa and the *Y*-axis or ordinates. The point of intersection of the two lines is called as the "origin". The measurements in the right side and above the origin are positive and that in the left and below the origin are negative. The plotting area is thus divided into four quarters known as quadrants. The quadrants are numbered in the anti-clockwise fashion starting from the quadrant one, which is above the *x*-axis and the right of *Y*-axis. However in many cases the graph may be drawn using one quadrant only. For example, if the values of *X* as well as *Y* are both positive then we may consider the first quadrant only.

(ii) **Scale Divisions:** In a coordinate system, the axes are used as scales for the representation of the numerical values. The scales are actually the spacing of units, whatever they happen to represent, is almost invariably different for the two axes. The scale divisions are indicated by small lines drawn perpendicular to the coordinate lines and are called as tic marks or scale points. Sometimes the tic marks are made to vary in size, accordingly the larger tic marks are termed as major tic mark and those smaller in size are called as minor tic mark. After frequent intervals the values represented by the various scale divisions should be marked with the corresponding numerical values.

(iii) **Coordinate Lines or Grid Lines:** The coordinate lines are important. But the grid lines should be kept as less as possible in the graph especially in cases where there are multiple number of curves in the graph. The gridlines if drawn should be lighter and thinner compared to the curves so that the curves appear prominent compared to the gridlines.

(iv) **Choosing of Scales:** The scales chosen for both the axis should be such that it results in a well proportioned chart. Such that the variation in the data of the resultant chart is neither minimized nor exaggerated.

(v) **Breaking the Scales:** Both the vertical scale or the horizontal scale may be broken depending on the purpose of the display. For example, it may sometimes so happen that only the upper portion of the coordinate field is required for visualization of the data and so the lower portion of the field may be eliminated. In such a case a break must be shown on the vertical scale at a point that will not interfere with the plotting of the data.

(vi) **Figures in the Scale:** The scales and scale figures run from left to right on the horizontal scale and from bottom to top on the vertical scale. The scale figures are kept close to the coordinate lines in such a way that they do not interfere with the plotted points or the curves.

(vii) **Axes Titles:** Usually both the axes (horizontal and vertical) carry not only figures but also explanatory titles to understand the type of data it represents. Such titles or levels also explain the units in which the data is measured along the axes. In cases where the figures in the scale are self explanatory the axes titles may be avoided. But in most cases since the response variable is represented along the *Y* axis so the title of that axis is almost mandatory.

(viii) **Legends:** When there is multiple numbers of curves or bars in the same graph representing different datasets, then it is essential to identify which curve or bar is used for which data set. Curves of different patterns or colors are used for the representation of the various variables and accordingly they are indexed in a box that is placed somewhere near or inside the chart without interfering the curve or the plotted points.

(ix) **Data Source and Data Labels:** The data source means the reference from which the data is collected. It is necessary to quote the report or book or journal from which the data, based on which the chart drawn, is collected. Data labels on the other hand means the numerical values of the data. The numerical values are noted down in the graph at close proximity to the point where it is represented in the graph. The data levels are generally omitted from the graph in order to make it neat, but obviously the presence of the data labels makes it easier to understand.

(x) **Title:** The final but most important attachment to a chart is its title. The chart title in general provides a declaration about the type of graph, the response variable that is represented and the independent variable on which the response variable is dependent. In case of time series data the title also declares the period of the graph. The title of a graph is generally placed at the top of the chart with a higher font size then the other text and has a center alignment.

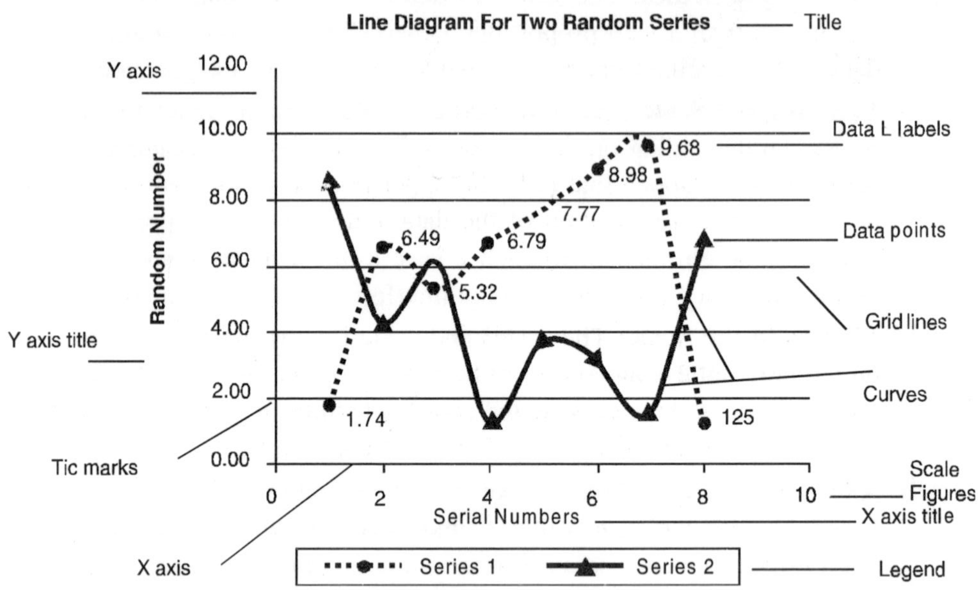

Figure 2.1: Parts of a Graph

Source: Data Generated in MS Excel

2.4 SOME TERMS RELATED TO DISTORTION IN CHART

(i) Data Density: The data density is an absolute measure. The data density is defined as the number of entries in the data set divided by the area of data graphic. The measure was proposed by Tufte (1983).

$$\text{Data Density} = \frac{\text{Number of entries in the data matrix}}{\text{Area of the data graphic}}$$

(ii) Data Ink Ratio: An effective way of avoiding clustering and over-plotting of the graph is by reducing the non-data parts of the display like gridlines, data labels, shading etc. The data ink ratio is based on this concept. It is also proposed by Tufte. The non-erasable part of a graphic and the non-redundant ink arranged, in response to variation in the numbers represented. The data ink ratio is defined as the ratio between the amount of ink used for the representation of the numerical values to the total ink involved in the graphic. A good graphic will have a high value of the data ink ratio and yet it remains easy to understand

$$\text{Data Ink Ratio} = \frac{\text{'ink' used to display tha data}}{\text{total 'ink' used to display the graphic}}$$

The data-ink ratio should be relatively high. A low data-ink ratio can be a result of 'chartjunk'. Types of "chartjunk" include

- unnecessary and distracting patterns,
- obtrusive grids, and
- graphics emphasizing style over quantitative information.

(iii) Lie Factor: The lie factor if a measure of the distortion in a graph. The lie factor in quantitative graphic is the maximum discrepancy between any quantity the graphic appears to display and its actual value. The best one can do is 1.0, larger or less the lie factor in a graph more is the distortion in the graphic.

$$\text{lie factor} = \frac{\text{Size of effect shown in graph}}{\text{size of effect in data}}$$

where

$$\text{size of effect} = \frac{|\text{second value} - \text{first value}}{\text{first value}}$$

A lie factor that is either much higher or much lower than one is bad. A high lie factor exaggerates differences between values. A low lie factor obscures differences between values.

2.5 CONCLUSION

There are many other terms that are associated with graphics. Discussion of the entire lot will definitely lead to an extended glossary. However, if we come across new terms associated with graphics in subsequent section of the text then we can provide footnotes and accordingly clarify the discussion.

Chapter 3

Some Simple But Less Common Graphs

3.1 INTRODUCTION

The most commonly seen graphs are the pie diagram, the bar diagram of different types, the line diagram etc. These diagrams have different purpose but are commonly used for data representation. Other than these commonly seen tools of data visualization, a there are a host other tools which can be used to provide a visual edge to several types of data. In this chapter an attempt has been made to place before the reader some of these techniques. These techniques can be used for data representation only but the use of these tools may give a new look to your data.

3.2 COLUMN PLOT (CIRCULAR BASE)

3.2.1 Definition and Description

This plot is similar to the multiple bar diagram but the base of the bars in no longer *X* axis but a circle from which the bars come out in different directions. Being placed in the circular base with the bars extending radially, the bars take a conic shape. The main purpose of the graph is to visualize the performance of different variables under different categories. The main problem with multiple bar diagram is that with increase in the number of categories the bars keep on increasing and it often becomes difficult to space all the bars along the *X* axis. Here, each observation variable is represented by a bar that comes out from the circumference of a small circle, such that the lengths of the bars are proportional to the value it represents. The central point is sometimes expanded to form a small circle. Several concentric circles are drawn that acts as gridlines for the plot. These bars are equally spaced around the central point. The different bars representing different variables may be differently colored under each category. Such use of color makes it easier to identify a particular variable in a particular category.

3.2.2 Working Data

The data for this purpose and the plot is taken from the website www.mediaware-infotech.com. The data relates to the Rate of inflation of advertisement and circulation growth of various Hindi dailies of India.

Table 3.1: Rate of Advertisement Inflation and Circulation growth in Hindi dailies

Newspaper	Ad. Infl. Rate	Circulation Growth
Amar Ujala	6.83	26.68
Hindustan	20.6	9.67
Dainik Jagaran (MP)	13	N.A.
Nav Bharat Times	13.73	5.28
Navabharat	3.92	5.83
Dainik Bhaskar	10.39	9
Rajasthan Patrika	5.63	–2.46
Aj	13.28	21.75
Dainik Jagaran (UP)	12.4	11.0

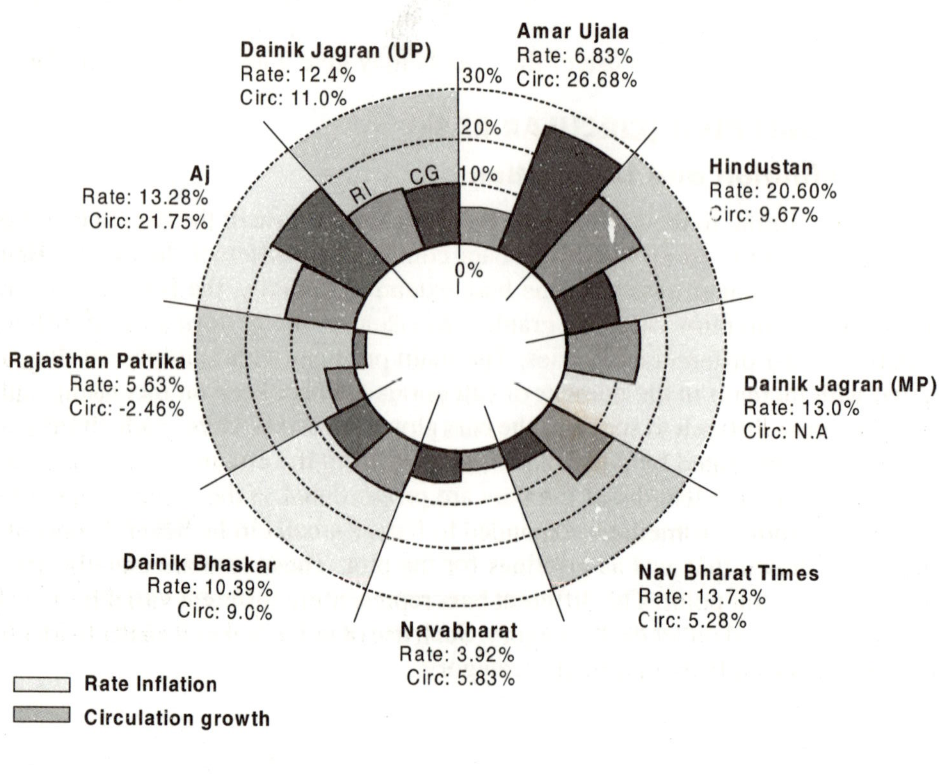

3.2.3 Axes

This plot does not require any axes for its drawing, only a radial scale is considered that can be demarked using concentric circles.

3.2.4 Advantages

(a) The plot is easy to interpret.

(b) The plot occupies less space compared to an equivalent multiple bar diagram.

3.2.5 Disadvantages

(a) The plot is difficult to draw manually.

(b) Most statistical software does not provide the option of drawing this plot.

(c) Large data sets rare difficult to represent.

3.3 DEVIATION PLOT

3.3.1 Definition and Description

There are different types of deviation plots and different software have used the name for plots completely different from each other. The deviation plot that is been discussed here may also be termed as the "bar deviation plot". In this plot the X axis generally comprise of various categories and the values of the response variable are plotted along the Y-axis. The individual data points are then represented by vertical bars. However, the bars connect the data points to a user-selectable baseline (parallel to X-axis). If the baseline value is different than the plot's Y-axis minimum, then individual bars will extend either up or down, depending on the direction of the "deviation" of individual data points from the baseline. Ideally if one considers the Y-axis as the base line then the plot will look almost like a bar diagram. However it may be recommended that the average value of Y may be used to draw a base line and accordingly the values above and below the averages can be identified.

3.3.2 Working Data

The data used for the plot is obtained from a survey conducted on the internet users of a town. 80 people were surveyed and their favorite Internet Café was noted. Accordingly the following data was obtained.

Table 3.2: Favorite internet café of the town

Name of the Café	No. of Respondents
Cyber Max	22
A to Z	20
Cyber World	17
Chit-o-Chat	10
Impact Graphics	7
Net Zone	4
Total	**80**

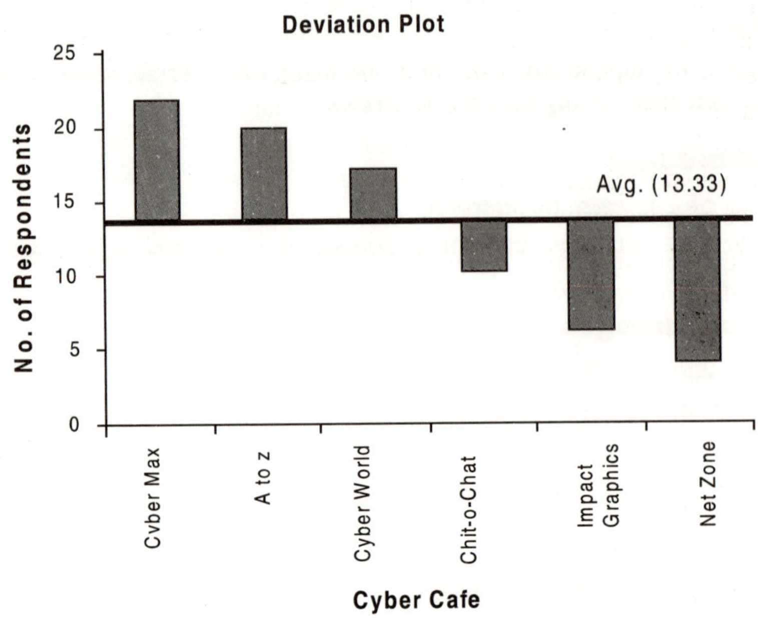

Figure 3.2: A Deviation plot for the data in Table 3.2

3.3.3 Axes

***X*-Axis:** It represents the various categories corresponding to which the values are provided. Here, various Internet Café is considered as the categories.

***Y*-Axis:** The vertical axis consists of that variable which is considered here as the response variable. For this case we took the number of respondents along the *Y* axis.

3.3.4 Uses

(a) The diagram can be used for data representation especially in cases where the magnitudes of one or more variables under different categories (or classes) are to be compared.

(b) It can also be used for the visualization of the mean of a set of observations.

(c) To understand which categories are performing below and which are above average.

3.4 FISH BONE CHART

3.4.1 Definition and Description

This technique is a semi-graphical tool used for analyzing complex problems which seem to have numerous and interrelated causes. It explores the relationship between the problem and its causes (by category) and presents them visually. It may also be called an Ishikawa Diagram, named after the Japanese Professor Kaoru Ishikawa, who first used this technique in 1943.

At the initial stage define the goal. Write the goal on the far right side of your diagram and draw a long horizontal arrow pointing towards it. This line may be considered as the backbone of the plot and leads to the goal. From the horizontal arrow, the user may be able to branch off some major and minor steps or problems that may be faced to reach the goal. First one must identify the most significant causes, or potential causes for reaching the goal. These will form the main branches from the "backbone arrow".

For detailed analysis, and drawing more sub-problems one may draw sub-branches from the branches and point out more problems.

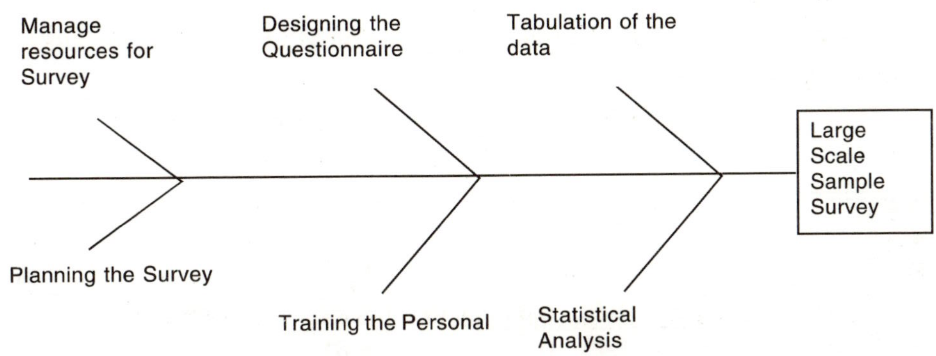

Figure 3.3: A Fishbone chart showing the steps in a sample survey

3.4.2 Uses

(a) It diagram can be used to understand the present position of a project and what is the next immediate target.

(b) It can be adapted for structuring a brainstorming session.

(c) It different surveys the technique can be used in explaining the enumerators and other officials associated with the survey about potential or actual problems and in overcoming the factors responsible for these causes.

3.5 HILO DIAGRAM

3.5.1 Definition and Description

The chart is used for the display of the highest and the lowest values of different data sets. Corresponding to each data set a vertical line is drawn parallel to the Y axis. The line starts from the minimum value and ends at the maximum value. Some small horizontal lines may be drawn from the vertical lines to signify the values of the mean, median etc. The different datasets are numbered 1, 2 etc. These integers are taken along the X axis and are used to identify the different datasets.

The plot can also be useful for share market experts to show the highest, lowest, closing as well as opening shares for several days in the same graph. For example:

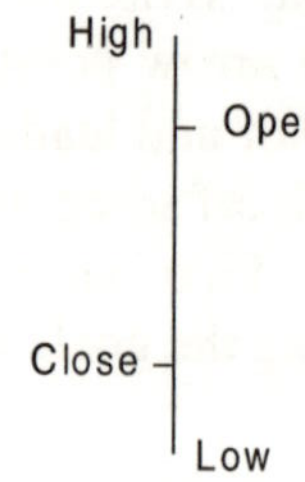

3.5.2 Working Data

The data used for drawing the diagram is generated in MS Excel using the function for generating random numbers.

Table 3.3: Different data sets generated in Excel

Data Set No.	Data Values						Highest	Lowest	Average
1	9.08	8.81	8.87	9.29	8.81	5.46	9.29	5.46	8.39
2	9.51	5.84	8.12	4.43	0.56	2.52	9.51	0.56	5.16
3	5.29	9.24	0.18	3.61	9.93	3.60	9.93	0.18	5.31
4	1.05	0.60	0.37	6.18	4.01	0.78	6.18	0.37	2.16
5	0.42	1.40	0.94	4.44	0.31	8.96	8.96	0.31	2.74
6	4.80	5.58	9.95	3.07	5.50	3.68	9.95	3.07	5.43

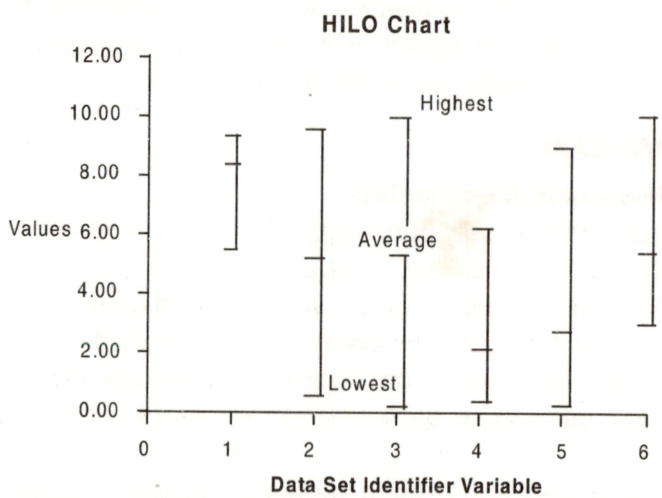

Figure 3.4: A HILO diagram corresponding to the data in Table 3.3

3.5.3 Axes

***X* Axis:** The axis is used to represent the data set identifier variable.

***Y* Axis:** This axis is used to represent the values of the response variable. Here we use the population along Y axis.

3.5.4 Uses

(a) The plot is simple to draw and the calculations are relatively easy.
(b) The chart is used to compare the means of several data sets.
(c) The chart is used for the purpose of comparison of the highest, lowest and the range of several data sets.
(d) The chart is useful in data representation in share markets.

3.6 RADAR PLOT

3.6.1 Definition and Description

This is a plot in the form of a circle. A circle is drawn first and the values of the independent variable are considered as an angle corresponding to a pre determined radius of the circle. The response variable is considered along the radius. The plotted points are then added and according a closed area across the radius is visible. Multiple numbers of response variables may also be plotted in the radar plot, producing several networks. In case of different response variables they may be differently colored for identification.

3.6.2 Working Data

The data provided in Table 3.4 is self explanatory and the plot below correspond the Radar plot to the data.

Table 3.4: Index of Industrial Production: Sector-wise (Base Year : 1993 – 94 = 100)

Industry	1997-98	1998-99
Mining	126.4	125.4
Manufacturing	142.5	148.8
Food Products	133.8	134.7
Tobacco	158.1	178.5
Cotton textiles	125.6	115.9
Wool, Silk	172	176.8
Jute	114.3	106
Textile	158.7	153.1
Wood	128.5	121.0
Paper	146.4	169.8

Source: Economic Survey, 2000-01.

3.6.3 Axes

This plot does not require any axes for its drawing, only a radial scale is considered that can be demarked using concentric circles.

3.6.4 Advantages

(a) The plot is easy to interpret.

(b) The plot occupies less space compared to an equivalent multiple bar diagram or multiple line diagrams.

(c) The option of drawing this plot is also available in MS Excel.

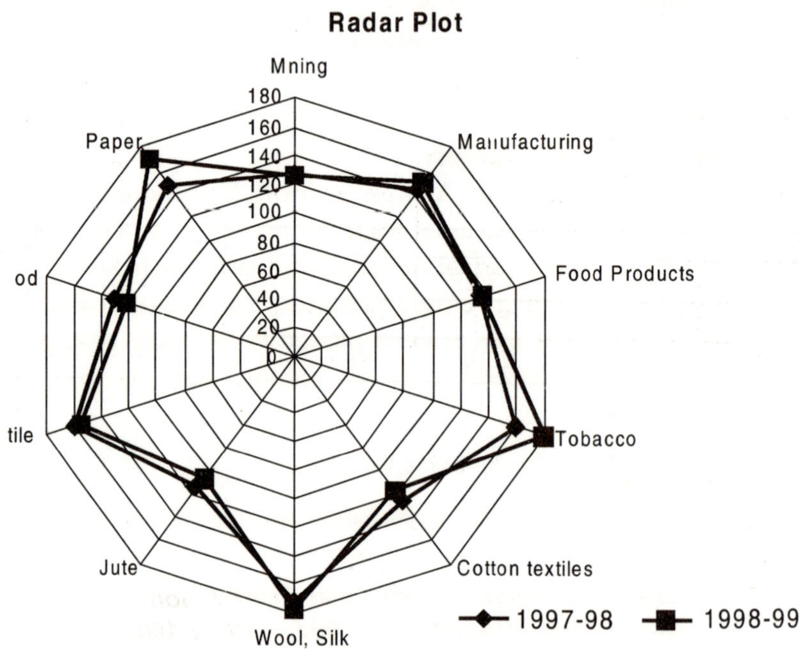

Figure 3.5: Radar Chart for the data in Table 3.4

3.6.5 Disadvantages

(a) The plot is difficult to draw manually.

(b) Variation of the response variable corresponding to a particular category may be difficult to perform

(c) Large data sets are difficult to represent.

3.7 STACKED LINE CHART

3.7.1 Definition and Description

This is a multiple line chart. But here each value of a particular response variable is obtained by adding with the previous one, in other words the cumulative totals of the different response variables are obtained. The values of the first response variables are plotted and then the plotted points are joined by straight lines. The values of the second response variable are then added to the first and the plotting is done. This process is repeated for each of the response variable. However it should be sensible to add together the data sets in to form the cumulative totals. It is obvious that the data should be of same unit.

3.7.2 Working Data

The data used for this plot is taken from Economic Survey of India, 1999-2001. The table shows the actual stock of food grains a hypothetical one representing data related to the production of an industry for different years.

Table 3.5: Actual stock of wheat in million tons.

	1999 (2)	2000 (3)	Cumu. Tot. (2)+(3)	2001 (5)	Cumu. Tot. (2)+(3)+(5)
January	12.7	17.2	29.9	25.0	54.9
April	9.7	13.2	22.9	21.5	44.4
July	22.5	27.8	50.3	38.9	89.2
October	20.3	26.9	47.2	36.8	84

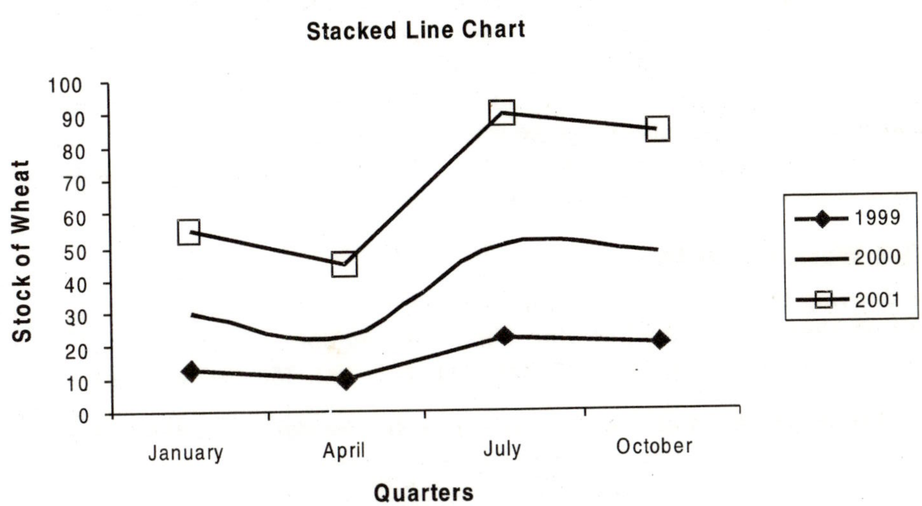

Figure 3.6: Stacked Line Chart based on data in Table 3.5

3.7.3 Axes

***X* Axis:** The axis is used to represent the independent variable. Here different quarters are used.

***Y* Axis:** This axis is used to represent the values of the response variable. The variable is Stock of Wheat in this case.

3.7.4 Advantages

(a) The plot provides a visual check related to the consistent variation in the data set.

(b) The lines do not cross each other, so the response variables can be studied easily.

(c) The graph can be drawn even without the use of color.

3.7.5 Disadvantages

(a) The plot cannot be used for the purpose of comparison of different points of the graph as there is no stable base line.

(b) The central values of the data set cannot be compared from this plot.

3.8 STEM AND LEAF DIAGRAM

3.8.1 Definition and Description

A stem and leaf plot, or stem plot, is a technique used to represent either *discrete* or *continuous* variables in a semi graphical display. A stem and leaf plot is used to organize data at the time they are collected and reduce the time of noting down all the numerical values. A stem and leaf plot looks something like a bar graph extended along the horizontal axis. Each number in the data is broken down into a stem and a leaf, thus the name. The stem of the number includes all except the last digit. The leaf of the number will always be a single digit. In case of numbers after decimal points it may be rounded of if the user thinks so. For drawing the diagram for a set of numerical values, on the left hand side of the page, write down the thousands, hundreds or tens (all digits except the last one). These are called as the stems. A line is drawn to the right of these stems. On the other side of the line, write down the last digit of a number. These are called as the leaves.

3.8.2 Working Data

For drawing the diagram let us consider the marks of 20 students taken from a class obtained from a total of 500. The marks are 356, 370, 419, 422, 333, 207, 350, 355, 372, 375, 378, 335, 337, 338, 411, 427, 293, 355, 297 and 295.

Table 3.6: A Stem and Leaf diagram to the data provided above.

Stem	Leaf
20	7
29	3 7 5
33	3 5 7 8
35	6 0 5 5
37	0 2 5 8
41	9 1
42	2 7

The leaves may be ordered in some cases i.e., in ascending or descending order.

3.8.3 Axes

The display does not require the use of any axes.

3.8.4 Uses

(a) It acts as a quicker way of noting down the values of the observation.

(b) It is better than a frequency table as it does not conceal the actual value of the observations.

(c) It can be used to understand the spread of data and is also useful in understanding the highest, lowest and the outliers.

Chapter 4

Graphs for Categorical Data

4.1 INTRODUCTION

Often while dealing with statistical data, one finds it easier (sometimes the only way) to divide the exhaustive set of observations into some non overlapping classes or categories. A variable used for the representation of such information is called as a categorical variable and the corresponding data as categorical data. The categorical data are generally represented in tabular form, with the category names along with their corresponding counts. Such a table gives the reader an idea about how frequently each of the categories occurs.

So far as visualization of categorical data are concerned the bar diagram causes a fairly nice representation. The bar diagram particularly is used for the display and comparison of the counts for each category, while the pie chart is used to show the contribution of each category to the whole group of cases.

However, when the same set of observations are to be categorized for two (or more) categorical variables the tabular representation takes the form of a two way table. Such a table, which is used for the display of two categorical variables, is called as contingency table. Since the table shows how the data distributed along each variable contingent in the value of the other variable, so is the name. The table given below is an example of a contingency table. The table is derived from a 5-way contingency table used in Edwards and Kreiner (1983). The data come from a sample of employed men aged in between 18 and 67, who were asked whether, in the preceding year, they had carried out any work on their homes.

Table 4.1: A Contingency Table Showing Age of Respondent and Labour Type

		Age	
Labour type	*Less than 30*	*31–45*	*Greater than 46*
Skilled	169	123	68
Unskilled	161	128	160
Office	187	345	250

4.2 ASSOCIATION PLOT

4.2.1 Definition and Description

The association plots originally proposed by Cohen (1980) visualize the cell frequency of a contingency table with the help of rectangles. But before drawing the plot one has to compute the expected values of the cell frequencies of the contingency table and accordingly compute the Pearsonian residuals.

For a two-way contingency table, the expected frequency of each cell, i.e. the frequency that a cell should have under the assumption that the two categorical variables (row and column variables) are independent can be easily computed. If e_{ij} denotes the expected frequency of the cell in the *i*th row and *j*th column, then we have $e_{ij} = \frac{r_i \times c_j}{n}$, where, r_i = *i*th row total, c_j = *j*th column total and n = total frequency.

Let, d_{ij} be the standardized residual corresponding to the *i*th row and *j*th column.

So, we have $d_{ij} = \frac{o_{ij} - e_{ij}}{\sqrt{e_{ij}}}$ where $i = 1, 2,..., r$ and $j = 1, 2,..., c$

Here, r is the total number of rows and c is the total number of columns. The value of d_{ij} is obtained for each cell.

Now, corresponding to each cell a rectangle is drawn. The rectangles in each row are drawn relative to a base line. The positive and negative values of the residuals are differentiated by rectangles growing above or below the base line respectively. The heights of the rectangles are drawn proportional to the corresponding Pearsonian residuals (d_{ij}) and the width are proportional to the square root of expected frequencies $\left(\sqrt{e_{ij}}\right)$.

We have,

length $\propto d_{ij}$ and width $\propto \sqrt{d_{ij}}$

i.e. length $\propto \dfrac{o_{ij} - e_{ij}}{\sqrt{e_{ij}}}$. Thus, Area = length × breadth $\propto \dfrac{o_{ij} - e_{ij}}{\sqrt{e_{ij}}} \times \sqrt{e_{ij}} = o_{ij} - e_{ij}$

So, area of each rectangle is proportional to the raw residual. Also, Cohen used different colors for positive and negative residuals. Mayer, Zeileis and Hornik (2003) proposed color and shading pattern similar to those used by Friendly (1994) for the extended mosaic plots can be used in association plot.

4.2.2 Working Data

The working data is provided in Table 4.1.

4.2.3 Advantages

(a) This plot is useful for visual representation of data arranged in the form of contingency tables.

(b) The plot is useful in identification of the cells in which the difference between observed and expected values is more.

4.2.4 Disadvantages

(a) The pattern of shading used in association plot is used to visualize significance but not the pattern of deviation from independence.

(b) The plot is not available in most statistical software.

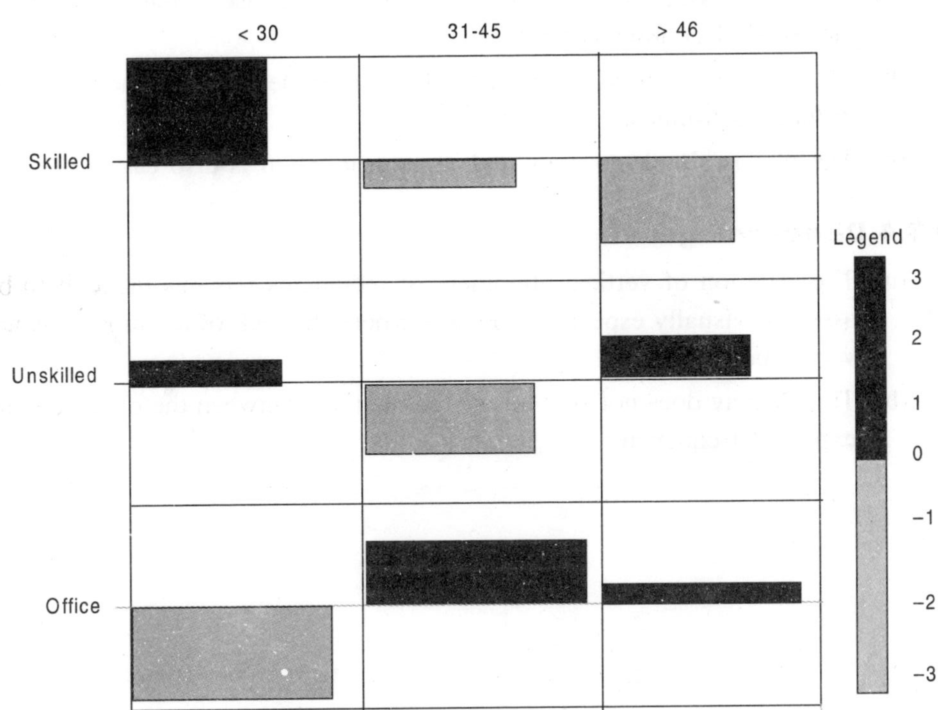

Figure 4.1: Cohen's Association Plot based on the data provided in Table 4.1

4.3 MOSAIC PLOT

4.3.1 Definition and Description

Mosaic plots are used for graphical representation of the two way or multi-way contingency tables. Mosaic plots can be stated as an extension of grouped bar-charts, where the width and height of the bars are proportional to the relative frequency of two variables.

The mosaic plot, proposed by Hartigan and Kleiner (1981) is basically a graphical method used to show the values of the contingency table which is cross classified by another categorical variable. In this diagram the cell frequency of the contingency table are represented by rectangular blocks or 'tiles' whose area is proportional to the cell frequencies provided in the table. Here each rectangle is drawn in such a way that the width of each rectangle is proportional to its column total and its height is proportional to the conditional frequency. Thus, the area of each tile is proportional to the cell frequency.

4.3.2 Working Data

The working data is provided in Table 4.1.

4.3.3 Advantages

(a The plot is used for the visualization of data in case of data arranged in the form of contingency table.

(b) The plot can be used to understand the dependence if any between two different attributes.

(c) The plot is simple to draw and its interpretation is also easy.

4.3.4 Disadvantages

(a) The criterion of vertical alignment of rectangles remains difficult to be assessed visually especially in cases where the lack of alignments is not very loud.

(b) This display does not support any comparison between the observed and expected frequencies.

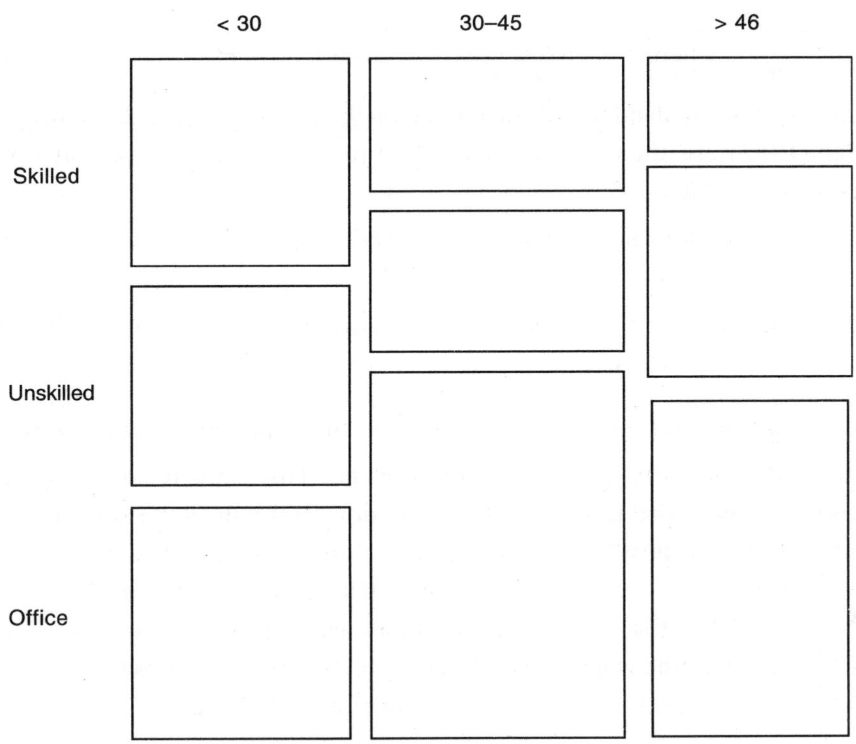

Figure 4.2: A Mosaic display as proposed by Hartigan and Kleiner (1981)

Here the area of each rectangle is proportional to the observed cell frequency.It may be noted that the tiles are separated slightly, in order to make small counts more visible and to make better visualization of the tiles for understanding, horizontal and vertical dimensions. The particular display can also be used to study the independence between two contingency variables. If the tiles of each row have the same height then it indicates that the two categorical variables are independent. In such a case, all the rectangles in each row are aligned vertically.

4.3.5 Friendly's Extension to Mosaic Display

Michael Friendly's works ushered a new era in the visual representation of categorical data. Friendly (1994) proposed some extension by the use of color which helps mosaic plots not only to display contingency tables but also visualizes the patterns of deviation from a specified model. Though the basic mosaic display proposed by Hartigan and Kleiner (1981) displays data in contingency tables but it does not in general provide a visual representation whether the data fits a specific model or not and also does not have any means to display the expected frequencies.

For a two-way contingency table, the expected frequency of each cell, i.e. the frequency that a cell should have under the assumption that the two categorical variables (row and column variables) are independent can be easily computed. If e_{ij} denotes the expected frequency of the cell in the ith row and jth column, then we have

$e_{ij} = \frac{r_i \times c_j}{n}$, where, r_i = ith row total, c_j = jth column total and n = total frequency.

First the mosaic display is drawn based on the proposals made by Hartigan and Kleiner (1981) are then colored using a color scheme based on standardized (Pearsonian) residuals.

Let, d_{ij} be the standardized residual corresponding to the ith row and jth column.

So, we have $d_{ij} = \frac{o_{ij} - e_{ij}}{\sqrt{e_{ij}}}$ $i = 1, 2,\ldots., r$ and $j = 1, 2,\ldots., c$

Here, r is the total number of rows and c is the total number of columns.

Now the value of d_{ij} is obtained for each cell. This value is used to determine the color to be provided to each of the rectangles. Under the color scheme used by Friendly, cells with positive deviations are colored blue and those with negative deviations are colored red. For $|d_{ij}| \leq 2$ the rectangles are empty and for $|d_{ij}| > 2$ the rectangles are filled, for $|d_{ij}| > 4$, the rectangles are filled with a darker pattern. For $|d_{ij}| \leq 1$ the dashed line is used to draw the boundary of rectangles and for $1 \leq |d_{ij}| \leq 2$ solid lines are used to draw the boundaries of the rectangles.

4.4 SIEVE'S DIAGRAM

4.4.1 Definition and Description

Sieve's diagrams are used for graphical representation of the two way contingency tables. Sieve's diagram is an extension over mosaic display originally developed by Hartigan and Kleiner (1981). Readers are requested to make a reading mosaic plot for better understanding of Sieve's diagram. Riedwyl and Schiipbach (1983) proposed a mosaic display in such a way that the area of the tiles in the mosaic is proportional to the expected frequencies. This diagram was later called as the 'Parquet Diagram'. In this display the height of each rectangle is proportional to its row total (r_i) and width of each rectangle is proportional to its column totals (c_j). Thus, the area is proportional to $r_i \times c_j$ which also is proportional to the expected frequency. Thus, we see that rectangles representing the values in each row have the same height. Similarly the rectangles representing values in each column have the same width and thus remain aligned horizontally.

4.4.2 Working Data

The working data is provided in Table 4.1.

4.4.3 Some Additional Insight

The basic plot does not have any scheme to display the difference between the observed and expected frequencies. Riedwyl and Schiipbach (1994) attached this by a modification to the plot. The modification was to divide each tile into small squares. The squares in each rectangle are equal to the observed frequencies. This helps, one to understand the difference between the observed and expected frequency, which appears from the density of shading. Also color was used to indicate whether the deviations of the observed frequency (o_{ij}) from the expected (e_{ij}) are either positive or negative. For example, negative residuals may be colored red and positive residuals blue. Thus, this display visualizes the observed frequency, expected frequency and the difference between the two.

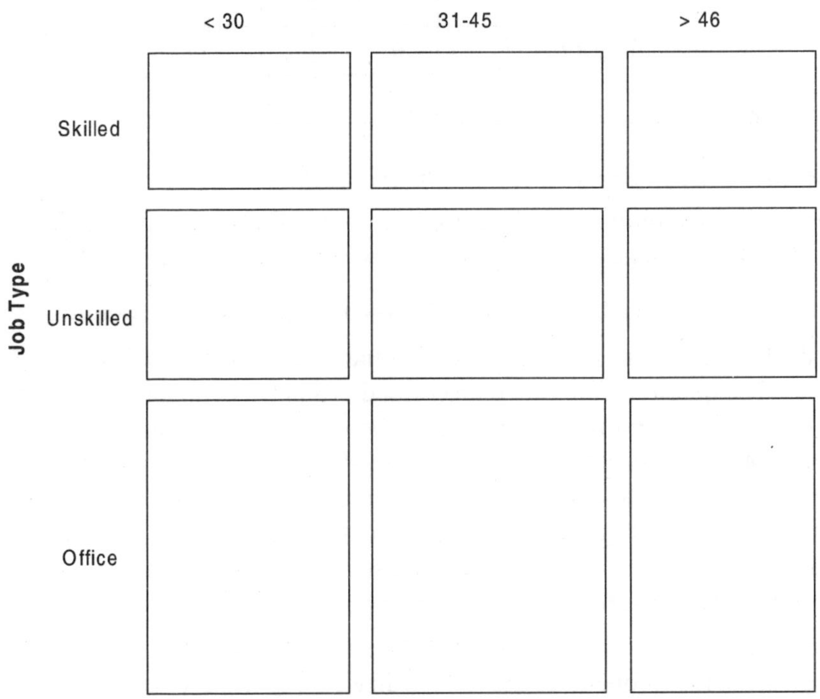

Figure 4.3: A Sieve diagram for data in Table 4.1

Here the area of each rectangle is proportional to the expected cell frequencies.

4.4.4 Advantages

(a) The plot is used for the visualization of expected frequencies of various cells in case of data arranged in the form of contingency table.

(b) The plot is simple to draw and its interpretation is also easy.

4.4.5 Disadvantages

(a) The display does not support visualization of the observed cell frequencies.

(b) This display does not support any comparison between the observed and expected frequencies.

(c) Commonly used statistical software does not provide the option to draw this plot.

4.5 FOUR FOLD DISPLAY

4.5.1 Definition and Description

A four fold display is used to display the cell frequency in a 2 ×2 contingency table. This was introduced by Friendly (1994b, 1994c). The display is done in such a way that it also displays the odds ratio of the contingency table. A 2 × 2 contingency table is generally in the following form:

		Category A	
		I	II
Category B	I	θ_{11}	θ_{12}
	II	θ_{21}	θ_{22}

The odds ratio is given by $\theta = (\theta_{11}/\theta_{12}) / (\theta_{21}/\theta_{22})$ (4.1)

If $\theta = 1$, then it implies that the categories A and B are mutually independent. In this display the frequency in each cell is shown by a quarter circle, whose radius is proportional to $\sqrt{o_{ij}}$, (where o_{ij} is the observed frequency corresponding to the *i*th row and *j*th column) so the area of each quarter is proportional to the observed frequency. The quarter circles are then colored or shaded depending on the direction and color of the Pearsonian residual. If there is an independence between the categorical variables then the odds–ratio (θ) tends to one, and the diagonally opposite quarters of the four fold display are approximately of the same size as that of the quarters in the other diagonal. However in case of association between variables the odd ratio is far from 1 and in such a case the diagonally opposite cells in one direction differs in size from those of the opposite direction.

However, some of the four fold displays the raw frequencies are not used for plotting instead the frequencies are standardized using iterative proportional fitting. This is done in order to make the quarter circles comparable.

4.5.2 Working Data

The table given below is a 2 × 2 table, from Walker and Lanes (1991). The table shows death or hospitalization due to asthma cross classified by Fenoterol use or not.

Table 4.2: Asthma deaths, and Fenoterol use from Walker and Lanes (1991)

Category 1 / Category 2	Fenoterol Used	Fenoterol Not Used
Deaths	34	50
Hospitalized	151	213

4.5.3 Advantages

(a) The plot is used for the visualization of data in case of data arranged in the form of contingency table.

(b) The plot can be used to understand the dependence if any between two different attributes.

(c) The plot is simple to draw and its interpretation is also easy.

4.5.4 Disadvantages

(a) The criterion of independence is sometimes difficult to be assessed visually and the user in all cases may not be confident in his decision.

(b) This display does not support any comparison between the observed and expected frequencies.

(c) The display is restricted to 2×2 contingency table only.

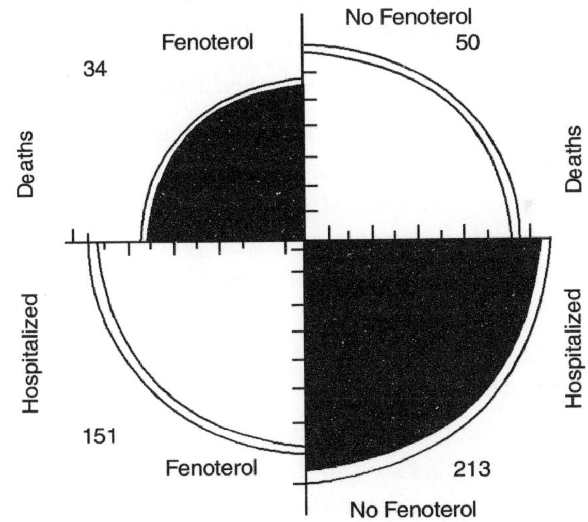

Figure 4.4: A four fold display for the raw frequencies given in Table 4.2

From the display we can see that the diagonally opposite cells in one direction show some difference in size in the opposite direction. This indicates that there is some association between the two categories.

REFERENCES

Cohen, A. (1980). On the Graphical Display of the Significant Components in a Two-Way Contingency Table. *Communications in Statistics—Theory and Methods,* A9, 1025–1041.

Edwards, D.E. and Kreiner, S. (1983). The analysis of contingency tables by graphical models. *Biometrika,* 70, 553–565.

Friendly, M. (1994a). Mosaic display for multiway contingency tables. *Journal of the American Statistical Association*, Vol. 89, No. 425, March 1994 Theory and Methods, 190–200.

Friendly, M. (1994b). A fourfold display for 2 by 2 by *K* tables. Technical Report 217, York University, Psychology Dept.

Friendly, M. (1994c). SAS/IML graphics for fourfold displays. *Observations*, Vol. 3, No. 4, 47–56.

Hartigan, J. A. and Kleiner, B. (1981). Mosaics for contingency tables. In W. F. Eddy (ed.), *Computer Science and Statistics: Proceedings of the 13th Symposium on the Interface*, 268–273. New York, NY: Springer-Verlag.

Riedwyl, H. and Schiipbach, M. S. (1983). Graphisehe darstellung von kontingenztafeln. *Technical Report 12*, Institute for Mathematical Statistics, University of Bern, Bern, Switzardland.

Walker, A.M. and Lanes, S.F. (1991). Misclassification of Covariates. *Statistics in Medicine,* 10, 1181–1196, Table 1.

Chapter 5

Fitting Discrete Distributions

5.1 INTRODUCTION

In this chapter we discuss some graphical displays that can be used to assess if a particular data set follows a specified discrete distribution. Let us suppose that X is a discrete random variable, such that, we have,

$$p_k = P(X = k)$$

Let us consider that a random sample with 'N' observations. These 'N' observations can be classified into a frequency distribution in accordance to the values assumed by X. This is obtained by calculating the probability p_k and then multiplying it with N. This is called as the theoretical frequency or expected frequency. Based on the expected values and the observed values some visual displays are developed that can be used as diagnostic tools for checking if the data assumes to have come from a given distribution. In the subsequent portion of this section, we deal with some existing plotting techniques for this purpose.

5.2 BINOMIALNESS PLOT

5.2.1 Description of the Plot

Hoaglin (1980) suggested a plot for checking the goodness of fit of a binomial distribution. Let us suppose that $X \sim$ *Binomial* (n, p), then we have

$$p(x) = p[X = x] = {}^nC_x P^x (1-p)^{n-x}, \quad x = 0, 1, 2, \ldots, n$$

$$= {}^nC_x \left[\frac{p}{1-p}\right]^x (1-p)^n \tag{5.1}$$

Now,

$$\text{Expected frequency} = \eta_x = N.P(X = x) = N.\,{}^nC_x \left[\frac{p}{1-p}\right]^x (1-p)^n \tag{5.2}$$

Taking *log* of both the sides,

$$\log \eta_x = \log N + \log {}^nC_x + x \log \left[\frac{p}{1-p}\right] + n \log(1-p)$$

$$\Rightarrow \log \eta_x - \log {}^nC_x = \log N + x \log \left[\frac{p}{1-p}\right] + n \log(1-p)$$

$$\Rightarrow \log \left[\frac{\eta_x}{{}^nC_x}\right] = \log N + n \log(1-p) + x \log \left[\frac{p}{1-p}\right] \tag{5.3}$$

which is a linear function in terms of x with $\log \left(\frac{p}{1-p}\right)$ as the slope and $\log N + n \log(1-p)$ as the intercept. Here we have the count metameter function as $\phi(x) = \log \left[\frac{\eta_x}{{}^nC_x}\right]$.

For drawing the binomialness plot, we compute the value of $\log \left[\frac{O_x}{{}^nC_x}\right]$, where O_x denotes the values of the observed frequencies. The points $\left(x, \log\left[\frac{O_x}{{}^nC_x}\right]\right)$ are then plotted in a graph paper along with the points $\left(x, \log\left[\frac{\eta_x}{{}^nC_x}\right]\right)$, which falls in a straight line and accordingly these points are connected to get the straight line. In case the observed data follows a binomial distribution then, the plotted points i.e. $\left(x, \log\left[\frac{O_x}{{}^nC_x}\right]\right)$ lies in close proximity to the line.

5.2.2 Working Data

The data that is used for the plot is taken from David (1971), which gives the number of fishes that are caught in one trap. In order to test if the data is a good fit for the binomial distribution we calculate the expected frequency in the usual procedure and written in the third row of the Table 5.1.

Table 5.1: Fish Catch Data from David (1971)

Fish per Trap	0	1	2	3	4	5	6	7	8
Observed Frequency	1	2	11	20	29	23	10	3	1
Expected Frequency	1	3	11	22	27	22	11	3	0

5.2.3 Calculations and Figure

Since the value of the parameter is not provided so one can estimate the value of '*p*' the probability of success in the usual procedure by using the maximum likelihood estimator which is given by, $\hat{p}=\bar{x}/n \, where \, \bar{x}=\sum f_x o_x / \sum f_x$ and accordingly estimate the expected frequencies.

Table 5.2: Calculations related to binomialness plot for the above data

Counts(x)	*Observed Frequency* (O_x)	*Expected Frequency* (η_x)	$log\left[\frac{O_x}{^nC_x}\right]$	$log\left[\frac{\eta_x}{^nC_x}\right]$
0	1	1	0	0
1	2	3	–1.38629	–0.98083
2	11	11	–0.93431	–0.93431
3	20	22	–1.02962	–0.93431
4	29	27	–0.8812	–0.95266
5	23	22	–0.88986	–0.93431
6	10	11	–1.02962	–0.93431
7	3	3	–0.98083	–0.98083
8	1	0	0	

5.2.4 Axes

***X*-Axis:** It represents the value of the independent variable. In the figure we take the number of fishes per trap along the axis.

***Y*-Axis:** The vertical axis consists of the values of the count metameter function which is given by $\log\left[\frac{\eta_x}{^nC_x}\right]$.

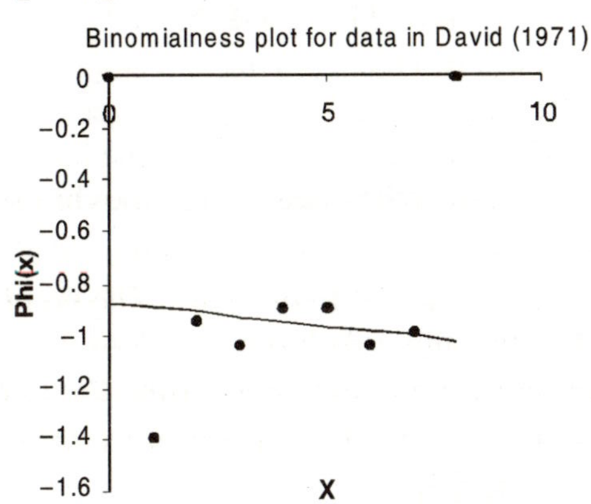

Figure 5.1: Binomialness plot for above data

Here we find that some of the points fall apart from the hypothetical line though most of the points are close proximity to the line. This indicates that there is an indication of lack of fit to a binomial distribution.

5.2.5 Uses

(a) This plot as the name indicates can be used for checking the goodness of fit of the data to a binomial distribution. The use is restricted and it is difficult to use the plot to decide about the lack of fit from looking at the binomialness plot. Some confidence bands around the observed line can be a step towards the greater acceptability of the technique.

(b) With a little modification the plot can be used to detect the goodness of fit of a data to a truncated binomial distribution.

(c) Probably none of the statistical software posses the plot by default.

5.3 POISSONNESS PLOT

5.3.1 Description of the Plot

This plot is used to check if a given data set can be considered to have come from a Poisson distribution. The plot was introduced by Hoaglin (1980). We know that if X follows Poisson distribution with parameter λ, then we have

$$P(X = x) = \frac{e^{-\lambda}\lambda^x}{x!}, \ \lambda > 0, x = 1, 2, 3, ...$$

Now, if the total frequency is N, then the expected frequencies of the distribution corresponding to $x = 0, 1, 2, ...$ etc. are

$$\eta_x = N.P(X = x) = N.\ \frac{e^{-\lambda}\lambda^x}{x!},$$

Taking the logarithm of both sides with respect to base 'e', we have

$$\log \eta_x = \log N - \lambda + x \log \lambda - \log (x!) \tag{5.4}$$

$$\Rightarrow \quad \log [\eta_x (x!)] = \log N - \lambda + x \log \lambda$$

Thus, from (5.4) it is clear that there is a linear relationship between x and $\log [\eta_x (x!)]$. Now, if $\log [\eta_x (x!)]$ is plotted against x then we get a straight line provided the data comes from a Poisson distribution. The fitted line has a slope of $\log \lambda$ and an intercept of $\log N - \lambda$. The slope of the fitted line can thus be used as an estimate of λ.

For drawing the poissonness plot, we compute the value of $\log[O_x (x!)]$, where o_x denotes the values of the observed frequencies. The points $(x, \log[O_x (x!)])$ are then plotted in a graph paper along with the line (5.4). In case the observed data follows a Poisson distribution then, the plotted points i.e. $(x, \log[O_x (x!)])$ lies in close proximity to the line (5.4). The value $\log[\eta_x(x!)]$ is called as the count metameter and is denoted by $\phi(\eta_x)$.

5.3.2 Working Data

The data used for this purpose is taken from Student (1906) and is provided in the Table 5.3 given below:

Table 5.3: Count and corresponding frequencies from Student (1906)

Count :	0	1	2	3	4	5	6
Frequency :	103	143	98	42	8	4	2

5.3.3 Calculations and Figure

The corresponding table i.e. Table 5.4 shows the relevant calculations required for a Poissonness plot and the Figure 5.2 is the corresponding Poissonness plot to the data which shows that the data is a fairly good fit to the Poisson distribution.

Table 5.4: Calculation of expected frequencies, observed and expected

Counts (x)	Observed Frequency (o_x)	Expected Frequency (η_x)	$\log[\eta_x(x!)]$	$\log[o_x(x!)]$
0	103	106	2.025306	2.012837
1	143	141	2.149219	2.155336
2	98	93	2.269513	2.292256
3	42	41	2.390935	2.401401
4	8	14	2.526339	2.283301
5	4	4	2.681241	2.681241
6	2	1	2.857332	3.158362
	$\Sigma o_x = 400$	$\Sigma \eta_x = 400$		

Count metameter values based on student (1906) data

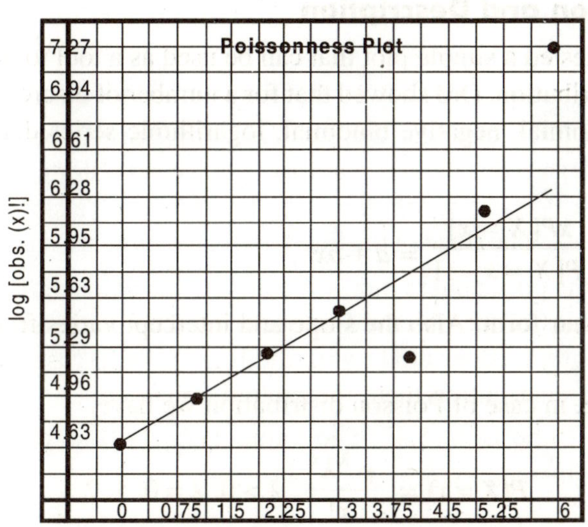

Figure 5.2: A Poissonness plot for the above data

In this plot the dots represents the values of $\log[O_x(x!)]$ *corresponding to the values of x. The straight line is* $\log[\eta_x(x!)] = \log N - \lambda + x \log \lambda$. *The closer the dots are to the line the better is the fit. However, the right tail of the straight line is very sensitive, small departure in the expected and observed counts at the end shows large departure in the graph. Since Poisson distribution deals with rare events so the value of λ will in general be small so the observed frequencies will be handsome for the lower values of x but it decreases considerably as the value of x increases. This makes the plot sensitive at the right tail.*

5.3.4 Axes

***X*-Axis:** It represents the value of the independent variable. In the figure we take the counts along the axis.

***Y*-Axis:** The vertical axis consists of the values of the count metameter function which is given by $\log [\eta_x (x!]$.

5.3.5 Uses

(a) This plot as the name indicates can be used for checking the goodness of fit of the data to a Poisson distribution. The use is restricted and it is difficult to use the plot in deciding about the lack of fit by just looking at the Poissonness plot. Some confidence bands around the observed line can be a step towards the greater acceptability of the technique.

(b) With a little modification the plot can be used to detect the goodness of fit of a data to a truncated Poisson distribution.

(c) Probably none of the statistical software posses the plot by default.

5.4 ORD PLOT

5.4.1 Definition and Description

Ord (1967) suggested a simple plot that can be used as a tool to diagnose the form of a discrete distribution. Ord showed that for a number of discrete distribution like the Poisson, binomial, negative binomial, logarithmic series distribution etc. the expression

$$\frac{xP[X = x]}{P[X = x - 1]} = a + bx$$

i.e of the linear form. Also the slope and intercept varies from distribution to distribution.

For example, in case of Poisson distribution we have,

$$P(X = x) = \frac{e^{-\lambda}\lambda^x}{x!}, \lambda > 0, \lambda = 0, 1, 2, ...$$

Thus, $$\frac{xP[X = x]}{P[X = x-1]} = \lambda \tag{5.5}$$

In case of binomial distribution, we have

$$P(X = x) = {}^nC_x\, p^x(1-p)^{n-x}, \quad x = 0, 1, 2,...,n$$

$$\Rightarrow \quad \frac{P[X = x]}{P[X = x-1]} = \frac{n-x+1}{x} - \frac{p}{1-p}$$

Thus, $$\frac{xP[X = x]}{P[X = x-1]} = \frac{(n+1)p}{1-p} - \frac{p}{1-p}x \tag{5.6}$$

For negative binomial distribution, we have

$$P(X = x) = {}^{x+r-1}C_x P^r q^x, \quad x = 0, 1, 2,..., p + q = 1$$

$$\Rightarrow \quad \frac{P[X = x]}{P[X = x-1]} = \frac{x+r-1}{x}q$$

Thus, $$\frac{xP[X = x]}{P[X = x-1]} = (r-1)\,q + qx \tag{5.7}$$

In case of the logarithmic series distribution, we have

$$P(X = x) = -\frac{\theta^x}{x\log(1-\theta)}, \quad 0 < \theta < 1, x = 0, 1, 2,...$$

$$\Rightarrow \quad \frac{P[X = x]}{P[X = x-1]} = \frac{-\theta^x}{x\log(1-\theta)} \times \frac{(x-1)\log(1-\theta)}{-\theta^{x-1}} = \frac{x-1}{x}\theta$$

Thus, $$\frac{xP[X = x]}{P[X = x-1]} = -\theta + \theta x \tag{5.8}$$

For the geometric distribution, we have

$p\,(X = x) = q^x p,\ 0 < p < 1,\ x = 0, 1, 2....$

$$\Rightarrow \quad \frac{P[X = x]}{P[X = x-1]} = q$$

Thus, $$\frac{xP[X = x]}{P[X = x-1]} = qx \tag{5.9}$$

In case of the discrete uniform distribution, we have

$$P(X = x) = \frac{1}{n+1}, \quad x = 0, 1, 2,\ldots,n$$

$$\Rightarrow \quad \frac{P[X = x]}{P[X = x-1]} = \frac{1/(n+1)}{1/(n+1)} = 1$$

Thus, $$xP[X = x]/P[X = x-1] = x \tag{5.10}$$

So, we find that from each of the equations (5.5), (5.6), (5.7), (5.8), (5.9) and (5.10) the form of the expression is a straight line but each of the straight line has a different form compared to the other in terms of their slopes and intercepts. The following table can explain it more clearly:

Table 5.5: Clue to understand the type of distribution from Ord Plots

Distribution	*Slope*	*Intercept*	*Estimate of the parameter*
Poisson (λ)	0	+	λ = *intercept*
Binomial (n, p)	–	+	p = *slope/(slope – 1)*
Neg. binomial (n, p)	+	+	p = *1 – slope*
Logaritmic Series distribution (θ)	+	–	θ = *slope*
Geometric (p)	+	0	p = *1 – slope*
Discrete uniform	+1	0	

5.4.2 Working Data

The data used for drawing the plot is taken from Jeffers (1978) which corresponds to the number of notices per day with its corresponding frequencies.

Table 5.6: Data from Jeffers (1978) along with expected frequencies from Poisson distribution

Notices per day	0	1	2	3	4	5	6	7	8	9
Observed Frequency	162	267	271	185	111	61	27	8	3	1
Expected Frequency	127	273	295	212	114	49	18	6	2	0

Table 5.7 below shows the calculations required for drawing the plot and deriving the fitted line.

Table 5.7: Calculations on data from Jeffres (1978) for Ord Plot

x	O_x	$P(X = x) = \frac{O_x}{\sum O_x}$	$y_x = \frac{xP[X = x]}{P[X = x-1]}$	*Estimated values of Y (Least Squares method)*
0	162	0.173448		1.6819
1	267	0.285867	1.648148	1.8253
2	271	0.29015	2.029963	1.9687
3	185	0.198073	2.04797	2.1121
4	111	0.118844	2.4	2.2555
5	61	0.06531	2.747748	2.3989
6	27	0.028908	2.655738	2.5423
7	8	0.008565	2.074074	2.6857
8	3	0.003212	3	2.8291
9	1	0.001071	3	2.9725

From the above table we have $\sum y = 21.60364$, $\sum x = 45$, $\sum xy = 116.64366$ and $\sum x^2 = 285$, we solve the normal equations and accordingly get the least square line as,

$$y = 1.6819 + 0.1434x \tag{5.11}$$

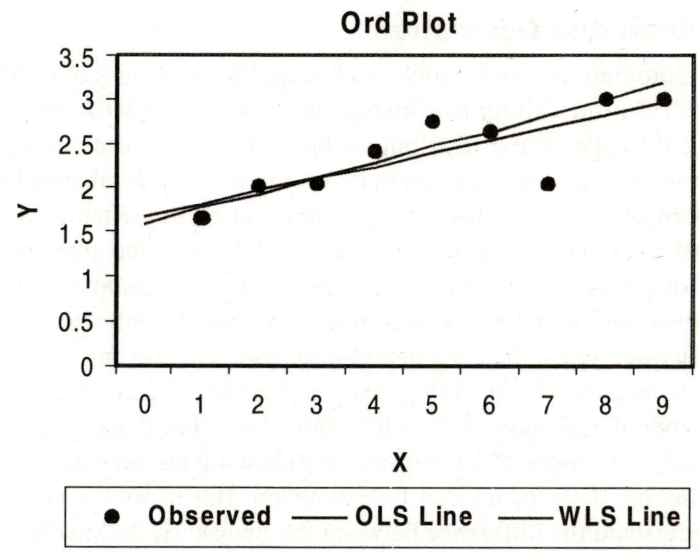

Figure 5.3: An Ord Plot for the data from Jeffres (1978) along with the fitted lines

The plot shows that the data can be considered to have come from a Poisson distribution. Both the lines WLS and OLS lines provide a reasonably good fit for the plots. The intercept of the OLS line is 1.5872 and that of WLS line is 1.6819 which are the estimates of λ under the OLS and WLS techniques respectively.

5.4.3 Axes

***X*-Axis:** It represents the value of the independent variable. In the figure we take the number of notices per day is considered along the axis.

***Y*-Axis:** The vertical axis consists of the values of the expression $y_x = xP[X=x]/P[X=x-1]$

5.4.4 Uses

(a) The plot can be used as a diagnostic tool for the distribution that best fits the data.

(b) The slope or intercept of the fitted straight line acts as a means to estimate the parameters of the distribution.

(c) The plot puts forward a visual impact on the quality of fit.

(d) The plot can be used to understand the pattern of variation between the observed and expected frequencies. For example, in the above figure one would find that the maximum variation for $X = 7$. Ordinary goodness of fit tests does not reveal such facts.

5.5 HANGING ROOTOGRAM

5.5.1 Definition and Description

The hanging rootogram is a useful tool for checking the goodness of fit of a data to a particular distribution. Though a histogram can be used for the purpose of understanding the type of the distribution but it has some limitations. First, the histograms are drawn in such a manner that the diagram gets dominated by the larger frequencies compared to the smaller frequencies that lie in the tail areas. Secondly, from the overall position of the plots one cannot decide about the inherent form of the distribution. For example, if the histogram is bell shaped we cannot conclude that it is from normal distribution as the normal distribution is not the only distribution having such a form. In order to evaluate the distribution more exactly, one must introduce a comparison between the observed frequency and the theoretical frequency obtained from the assumed distribution in the plot. Thus, one may think of a histogram of observed frequencies plotted along with a curve showing the expected frequencies in such a way that the discrepancy can be visualized. But in such a case it becomes difficult to understand the difference between the two as: (i) the smaller frequencies especially at the tails are difficult to visualize (ii) it becomes difficult to understand the difference between the histogram bars and the curve.

Keeping the two points in mind Tukey (1977) introduced the hanging rootogram. It is to root the frequencies and hang them to the theoretical curve and hence the name of the curve. In this plot, the following changes are made to a histogram:

(a) The frequencies are plotted on a square root scale, to make small frequencies relatively more prominent.

(b) The histogram bars start from the curve so that one may judge the differences more easily against a horizontal line.

5.5.2 Working Data

The data used for drawing the plot is taken from Jeffers (1978) which corresponds to the number of notices per day with its corresponding frequencies. The data is provided in Table 5.8. The data is fitted to a Poisson distribution and the corresponding expected frequencies are computed.

5.5.3 Axes

***X*-Axis:** It represents the value of the independent variable. In the figure we take the number of notices per day is considered along the axis.

***Y*-Axis:** The vertical axis consists of the values of the frequencies both observed and expected.

5.5.4 Uses

(a) The plot puts forward a visual impact on the quality of fit.

(b) The plot can be used to understand the pattern of variation between the observed and expected frequencies. For example, in the above figure if one looks at the base of the bars it appears that there is a periodic pattern noticed in the expected frequencies under the Poisson models. Ordinary goodness of fit tests does not reveal such facts.

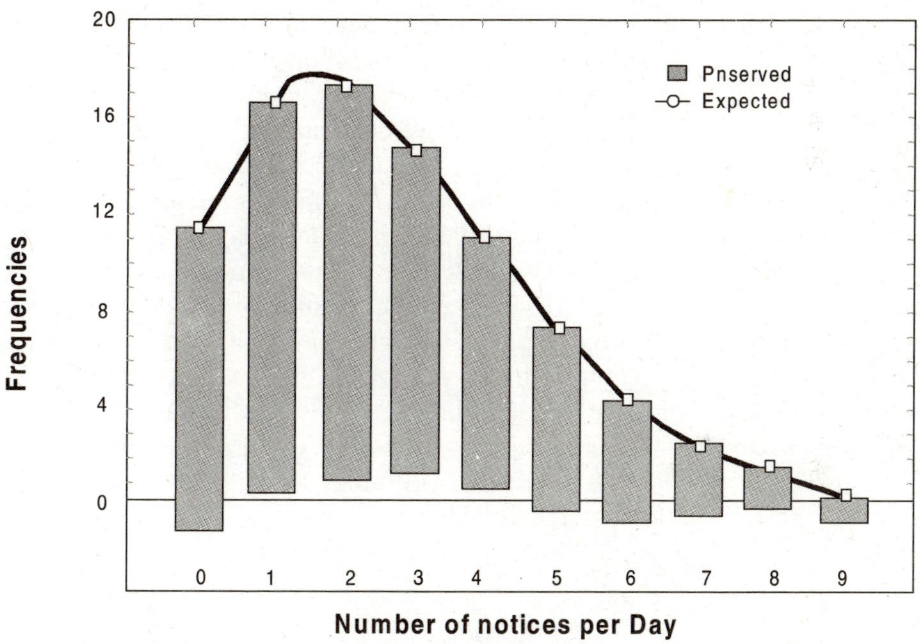

Figure 5.4: A hanging rootogram for the data in Jeffers (1978)

REFERENCES

David, F. N. (1971). *A First Course in Statistics*, 2nd Edition, Griffin.

Hoaglin, D.C. (1980). A Poissonness plot. *The American Statistician,* Volume 34, 3, 146–149.

Jeffers, J. N. R. (1978). *An Introduction to System Analysis with Ecological Applications*. London: Edward Arnold.

Ord, J. K. (1967). Graphical methods for a class of discrete distributions. *Journal of the Royal Statistical Society, Series A,* 130: 232–238.

Student (1906). On the error of counting with a haemocytometer. *Biometrika,* 5, 351–360.

Tukey, J. W. (1977). *Explanatory Data Analysis* (First Edition), Vol. 1, Ch. 5, Reading, Mass: Addison-Wesley Publishing Co.

Chapter 6

Fitting Continuous Distribution

6.1 INTRODUCTION

The testing of distributional patterns present in the data sets is often a matter of concern for statisticians. There are several techniques for fitting the assumed distribution and numerous parametric and non-parametric tests available for testing the goodness of fit. However, many graphical techniques are also available to check the goodness of fit to the hypothetical distribution under consideration. Literature of Kotz and Johnson (1988), Rice (1995) and others referred to such graphical techniques as informal techniques for data analysis and model assessment. But Fisher (1983), Doksum (1977) etc. described various graphical techniques, which can be used for formal statistical inference as well. Here we are going to discuss some of those graphical techniques that are used for checking the goodness of fit to continuous distributions either formally or informally.

6.2 PROBABILITY PLOT

6.2.1 Definition

Probability plot is a graphical technique for accessing whether sample data conform to a hypothesized distribution, based on subjective visual examination of the data. Our purpose is to test if the data follows a given theoretical distribution or not. The data is plotted against the theoretical distribution, in such a way that, the points should fall approximately in a straight line. If the plotted values show departure from the straight line then it implies that the data does not follow the specific distribution. The intercept and slope of the straight line are useful in the sense that they are the estimates of the location and scale parameters of the distribution. Though it is not so much important for the normal distribution, as the mean and standard deviation are the estimates of the location and the scale, it can be of great use in case of other distributions.

6.2.2 Description

Let $y_1, y_2, \ldots, y_n$ be a random sample of size n. We are to test if the random sample comes from the null distribution $F_0(y)$. This is the probability distribution of Y given by the null hypothesis. The steps for the construction of a probability plot are as follows:

(a) Sample is first ordered or arranged in increasing order of magnitude. The arranged sample is thus denoted by $y_{(1)}, y_{(2)}, \ldots, y_{(n)}$.

(b) The hypothetical probability $F_0(y_{(i)})$ for each $y_{(1)}, y_{(2)}, \ldots, y_{(n)}$ are calculated.

(c) We then apply continuity correction to the indicator variable i by subtracting ½ from each of them.

(d) The values $(F_0(y_{(i)}), i - \frac{1}{2})$ are then plotted in a graph sheet. Alternately one can plot $(y_{(i)}, \frac{i}{n+1})$. If the plotted points fall approximately in a straight line then it may be concluded that the random sample comes from the null distribution, $F_0(y)$, otherwise not.

6.2.3 Working Data

The data here is generated in MS Excel, is a set of random numbers (y_i, $i = 1, 2, \ldots, n$) from a uniform distribution ($U(0,1)$), which is then arranged to get the ordered sample ($y_{(i)}$, $i = 1, 2, \ldots, n$). The values of $(F_0(y_{(i)}), i - \frac{1}{2})$ and the expected frequencies are then calculated and tabulated.

Table 6.1: Data and calculations for probability plot

y_i	$y_{(i)}$	$F_0(y_{(i)}) = y_{(i)}$	i-(1/2)	Exp. freq
0.449568	0.048341	0.048341	0.5	0.96682
0.764733	0.102756	0.102756	1.5	2.05512
0.137883	0.118686	0.118686	2.5	2.37372
0.272835	0.137883	0.137883	3.5	2.75766

Cont...

y_i	$y_{(i)}$	$F_0(y_{(i)}) = y_{(i)}$	$i - (1/2)$	Exp. freq
0.973296	0.231117	0.231117	4.5	4.62234
0.450179	0.251228	0.251228	5.5	5.02456
0.102756	0.272835	0.272835	6.5	5.4567
0.941557	0.396435	0.396435	7.5	7.9287
0.396435	0.449568	0.449568	8.5	8.99136
0.535783	0.450179	0.450179	9.5	9.00358
0.853236	0.451827	0.451827	10.5	9.03654
0.451827	0.507218	0.507218	11.5	10.14436
0.231117	0.535783	0.535783	12.5	10.71566
0.983642	0.764733	0.764733	13.5	15.29466
0.507218	0.853236	0.853236	14.5	17.06472
0.936522	0.87701	0.87701	15.5	17.5402
0.118686	0.936522	0.936522	16.5	18.73044
0.251228	0.941557	0.941557	17.5	18.83114
0.048341	0.973296	0.973296	18.5	19.46592
0.87701	0.983642	0.983642	19.5	19.67284

6.2.4 Advantages

(a) Probability plots provide a quick check to the goodness of fit.

(b) Probability plots perform well both with large and small samples unlike other statistical tests which have strict restriction on sample sizes. Like the Chi-Square test for the goodness of fit test, which, require at least 50 to 100, observations for meaningful tests or the Wilks Shapiro test, which can be used only when the sample size lies between 10 to 1000.

(c) Probability plot can be used for a large variety of distributions.

(d) The intercept and slope of the straight line, which the plotted points approximately follow, can be used as the location and scale parameter of the distribution.

6.2.5 Disadvantage

The most striking problem with the plot is that different people reach to different conclusions regarding the goodness of fit test. Since, no proper confidence bands are developed around the straight line obtained from the plotting of expected frequency from the hypothetical distribution, so accessing the goodness of fit sometimes, becomes difficult even for experts.

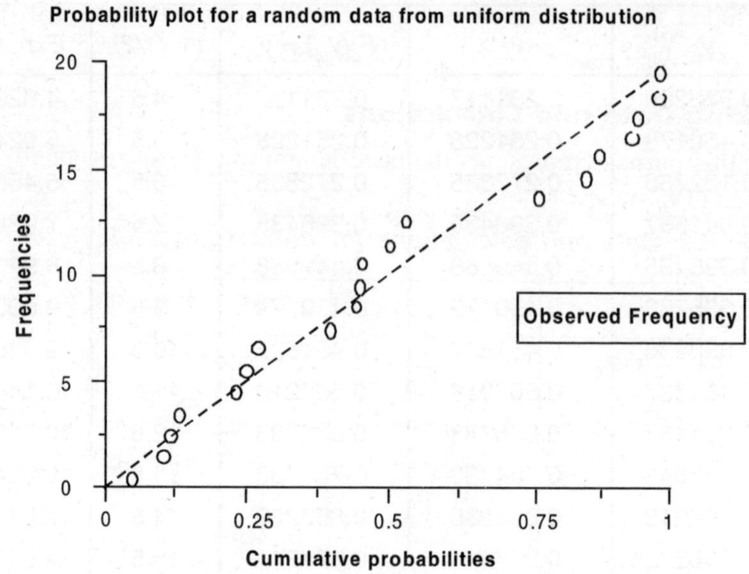

Figure 6.1: A probability plot for the data in Table 6.1

From the plot we see that the observed frequencies approximately follow the straight line and so we may conclude that the set of random numbers shows a nearly good fit. We then perform one sampled K-S test, and see that the data follows uniform distribution.

6.3 NORMAL PROBABILITY PLOT

6.3.1 Definition and Description

A substantial portion of inferential statistics is based on the assumption that the underlying distribution from which the sample is drawn follows a normal probability distribution. The normal probability plot is the probability plot for checking the normality of a data set. The normality assumption acts as a gatekeeper to many statistical methodologies relating to estimation, testing of hypothesis and the like. A quick way to assess the normality assumption can be performed by the normal probability plot.

Let $y_1, y_2, \ldots, y_n$ be a random sample of size n. We are to test if the random sample comes from a normal distribution. The steps for the construction of a probability plot are as follows:

(a) Sample is first ordered or arranged in increasing order of magnitude. The arranged sample is thus denoted by $y_{(1)}, y_{(2)}, \ldots, y_{(n)}$.

(b) For $i = 1, 2, \ldots, n$ the values of $(i - 0.5)/n$ are calculated.

(c) We then find $z_i = \phi^{-1}((i - 0.5)/n)$, where ϕ denotes the standard normal distribution function.

(d) The points $(y_{(i)}, z_i)$, $i = 1, 2, \ldots, n$ are then plotted in a graph sheet. If the plotted points fall approximately along the straight line then it may be concluded that the random sample comes from a normal distribution, otherwise not.

6.3.2 Working Data and Calculations

The data for this purpose represents the head length of 25 sons, originally taken from Anderson (1958).

Table 6.2: Data and calculations for normal probability plot

$y_{(i)}$	$(i–0.5)/n$	$z_i = \phi^{-1}((i–0.5)/n)$
163	0.0208	–2.0368
174	0.0625	–1.5341
174	0.1041	–1.2581
175	0.1458	–1.0544
176	0.1875	–0.8871
176	0.2291	–0.7415
179	0.2708	–0.6102
181	0.3125	–0.4887
181	0.3541	–0.3741
183	0.3958	–0.2641
183	0.4375	–0.1573
186	0.4791	–0.0522
188	0.5208	0.0522
188	0.5625	0.1573
189	0.6041	0.2641
190	0.6458	0.3740
191	0.6875	0.4887
192	0.7291	0.6102
192	0.7708	0.7415
195	0.8125	0.8871
195	0.8541	1.0544
197	0.8958	1.2581
197	0.9375	1.5341
208	0.9791	2.0368

6.3.3 Axes

***X* Axis:** The axis is used to represent the values of the observations.

***Y* Axis:** This axis is used to represent the Z values corresponding to each value of the observation.

6.3.4 A Parametric Test to Check the Normality of Data

The normal probability plot based on the data shows that the head lengths of the sons follow normal distribution. The result can also be conformed by using

Anderson-Darling test. The Anderson-Darling test is a modification of the K-S test as it gives more weight to the tails than the K-S test. It was also shown by Stephen (1974), that for a wide class of alternatives with unknown parameters, that Anderson-Darling test is typically more powerful than any other distance test for testing normality.

```
          ANDERSON-DARLING 1-SAMPLE TEST
          THAT THE DATA CAME FROM A NORMAL DISTRIBUTION
1. STATISICS:
   NUMBER OF OBSERVATIONS                     =        25
   MEAN                                       =  185.5417
   STANDARD DEVIATION                         =  9.930099
   ANDERSON-DARLING TEST STATISTIC VALUE      = 0.2348938
   ADJUSTED TEST STATISTIC VALUE              = 0.2638477
2. CRITICAL VALUES:
       95      %POINT   =   0.7350000
       99      %POINT   =    1.021000
1. CONCLUSION (AT THE 5% LEVEL):
     THE DATA DO COME FROM A NORMAL DISTRIBUTION.
```

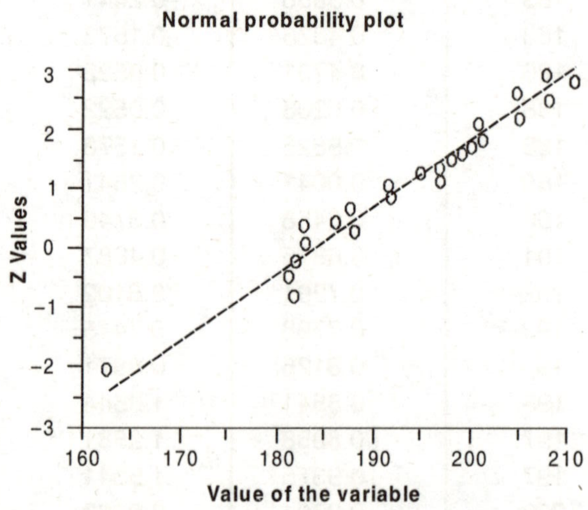

Figure 6.2: A normal probability plot to the data in Table 6.2

On looking at the figure it appears that the data is a good fit to normal distribution which conform to the result obtained from the results obtained from the Anderson-Darling test.

6.3.5 Advantages

(a) Normal Probability plots provide a quick check to normality of the data, which is often a necessary condition for many statistical tests.

(b) In regression or ANOVA, we often assume that the residual component is normally distributed. This plot can be used to check the validity of that.

6.3.6 Disadvantage

The problem with this type of plot is that the conclusions regarding the goodness of fit test taken by different users may be subjective. Since, no proper confidence bands are developed around the straight line obtained from the plotting of expected frequency from the normal distribution, so accessing the goodness of fit, may sometimes become difficult even for the experts.

6.4 EMPIRICAL DISTRIBUTION FUNCTION PLOT

6.4.1 Definition and Description

Let (y_i, $i=1,2,...,n$) be a random sample of size 'n' on Y, let $F(y)$ be the cumulative distribution function (CDF) of the random variable. Also, let ($y_{(i)}$, $i=1,2,...,n$) denotes the corresponding order statistics. The empirical CDF is given by, $F_n(y)$ = (no. of observations $\leq y$) / n

Thus this function will be a step function lying between 0 and 1, with jumps of size $1/n$ at each observation. This empirical distribution function popularly called as the EDF, is used as the non–parametric estimate of the actual CDF. The values of $F_n(y)$ are plotted corresponding to the values of y. The cumulative probabilities are computed based on the theoretical distribution that is to be fitted. If the parameters of the theoretical distribution are not known using the maximum likelihood estimator and the parameters can be obtained and accordingly the corresponding CDF curve is drawn. In case the data is a good fit to the corresponding theoretical distribution then the CDF curve and the EDF step polygon will lie in close proximity to each other otherwise they will be separated apart.

6.4.2 WORKING DATA AND CALCULATIONS

The data collected for the purpose is taken from Anderson (1958) which pertains the head length of a group of individuals. To check if the observations can be considered to have come from normal distribution we draw the EDF plot and compare it with the corresponding CDF. If the parameters of the normal distribution (μ, σ^2) are unknown we can estimate the parameters using their corresponding maximum likelihood estimators i.e. $\hat{\mu}=\bar{x}=\frac{1}{n}\sum x_i$ and $\hat{\sigma}^2=\frac{1}{n}\sum(x_i-\bar{x})^2$. The calculations for the data along with the expected frequencies for raw data from Anderson (1958) have been given below.

Table 6.3: EDF values for the Anderson Data

Head length	Expected F(y)	EDF Values	Head Length	Expected F(y)	EDF Values
163	0.00135	0	188	0.5	0.458333
174	0.00298	0	188	0.598706	0.5
174	0.00621	0	189	0.691462	0.666667

Contd....

175	0.012224	0.041667	190	0.773373	0.791667
176	0.02275	0.041667	191	0.841345	0.875
176	0.040059	0.041667	192	0.89435	0.958333
179	0.066807	0.041667	192	0.933193	0.958333
181	0.10565	0.041667	195	0.959941	0.958333
181	0.158655	0.166667	195	0.97725	0.958333
183	0.226627	0.166667	197	0.987776	0.958333
183	0.308538	0.291667	197	0.99379	1
186	0.401294	0.458333	208	0.99702	1

6.4.3 Advantages

(a) EDF plot provides a quick check to normality of the data, or of any other distribution that the data may follow.

(b) It is easy to draw and simple to understand.

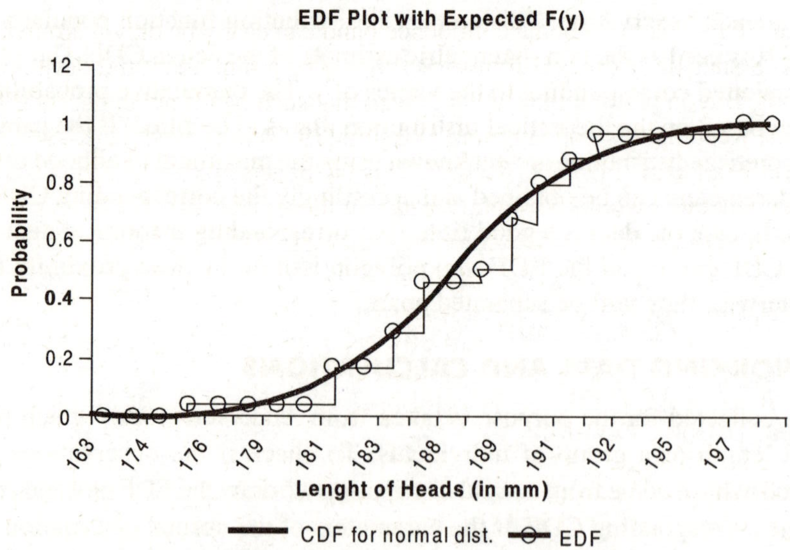

Figure 6.3: EDF plot for the Anderson Data

From the plot it appears that the data is a considerably good fit to the normal distribution.

6.4.4 Disadvantages

(a) The problem with this type of plot is that the conclusions regarding the goodness of fit test, taken by different users, may be subjective as the proximity of the EDF curve and the CDF curve may be differently interpreted by the users. A suggestion can be to develop a proper confidence bands around the CDF curve and so the step function lying within the confidence band may act as the indicator for a good fit.

(b) It is better if the theoretical distribution is fully specified along with the parameters, as the critical region of the Kolmogrov-Smirnov test may not remain valid if the parameters are estimated from the data.

(c) The plot remains scale sensitive i.e. by changing the scale along the *Y*-axis one can manipulate the conclusion related to the goodness of fit.

6.5 EMPIRICAL DISTRIBUTION FUNCTION PLOT (WITH DOKSUM BOUNDS)

6.5.1 Definition and Description

The plot is similar to that of the EDF plot discussed earlier, only that this step goes forward to generate a confidence band around the CDF curve such that the disadvantage of the EDF plot can be overcome.

We know that the famous non-parametric test, Kolmogrov – Smirnov test, that is used to test if a given random sample comes from a specific distribution, is based on the EDF and the test statistic is given by $D = Sup\ |F_n(y) - F(y)|$

If the CDF is plotted, then a confidence band can easily be drawn around it based on the K–S statistic. Kotz and Johnson (1988) give a hint about the development of such bands. But on plotting the band it is seen that such bands are unnecessarily broad in the tails. Keeping this issue in mind Doksum (1977) improved the K–S Statistic by dividing it by a factor $\{F(x)\,(1 - F(x))\}^{1/2}$. This factor acts as variance equalizer and the band thus generated would be slightly wider in the middle and much narrower at the tails compared to the first band. Accordingly the following expression may be derived.

$$1 - \alpha = P[Sup \frac{|F_n(y) - F(y)|}{[F(y)(1 - F(y))]^{1/2}} \leq D_\alpha(n)]$$

$$= P[\frac{|F_n(y)\ \ F(y)|}{[F(y)(1\ \ F(y))]^{1/2}} \leq D_\alpha(n)\ \forall y]$$

$$= P[-D_\alpha^{(n)} \leq \frac{F_n(y)\ \ F(y)}{[F(y)(1\ \ F(y))]^{1/2}} \leq D_\alpha(n)\ \forall y]$$

$$= P[F(y) - \sqrt{F(y)(1\ \ F(y)}\, D_\alpha(n) \leq F_n(y) \leq F(y) + \sqrt{F(y)(1\ \ F(y)}\, D_\alpha(n)\ \forall y]$$

Thus, $[F(y) - \sqrt{F(y)(1\ \ F(y)}\ D_\alpha(n),\ F(y) + \sqrt{F(y)(1\ \ F(y)}\, D_\alpha(n)]$ is the $100(1 - \alpha)\%$ Doksum confidence band for $F_n(y)$.

Thus for different values of the variable one may plot the EDF, CDF of the theoretical distribution and the Doksum bound deduced for a given level of significance. If the EDF lies within the band then the data may be considered as a good fit to the theoretical distribution.

6.5.2 Working Data and Calculations

The data collected for the purpose is taken from Anderson (1958) which pertains the head length of a group of individuals. To check if the observations can be considered to have come from normal distribution we draw the EDF plot and compare it with the corresponding CDF. It is better if the distribution is completely specified otherwise the parameters are estimated using their corresponding maximum likelihood estimators i.e. $\hat{\mu} = \bar{x} = \frac{1}{n}\sum x_i$ and $\hat{\sigma}^2 = \frac{1}{n}\sum (x_i - \bar{x})^2$. The calculations for the data along with the EDF values and the Doksum bounds for raw data from Anderson (1958) have been given below.

Table 6.4: EDF and Doksum bounds of the data

Head length	Expected F(y)	EDF	Lower Doksum Bounds	Upper Doksum Bounds
163	0.00135	0	–0.00856	0.011264
174	0.00298	0	–0.01174	0.017697
174	0.00621	0	–0.015	0.02742
175	0.012224	0.04167	–0.01744	0.041894
176	0.02275	0.04167	–0.01751	0.063009
176	0.040059	0.04167	–0.01289	0.093006
179	0.066807	0.04167	–0.00061	0.134223
181	0.10565	0.04167	0.022655	0.188645
181	0.158655	0.16667	0.06001	0.257301
183	0.226627	0.16667	0.113592	0.339663
183	0.308538	0.29167	0.183827	0.433248
186	0.401294	0.45833	0.26895	0.533637
188	0.5	0.45833	0.365	0.635
188	0.598706	0.5	0.466363	0.73105
189	0.691462	0.66667	0.566752	0.816173
190	0.773373	0.79167	0.660337	0.886408
191	0.841345	0.875	0.742699	0.93999
192	0.89435	0.95833	0.811355	0.977345
192	0.933193	0.95833	0.865777	1.000608
195	0.959941	0.95833	0.906994	1.012887
195	0.97725	0.95833	0.936991	1.017508
197	0.987776	0.95833	0.958106	1.017445
197	0.99379	1	0.97258	1.015001
208	0.99702	1	0.982303	1.011737

6.5.3 Advantages

(a) EDF plot provides a quick check to normality of the data, or of any other distribution that the data may follow at any required level of significance.

(b) It easy to draw and simple to understand.

(c) The use of bands overcomes the disadvantage related to the subjective approach of different users related to the interpretation of the plot.

(d) The plot does not remain scale sensitive.

6.5.4 Disadvantage

It is better if the theoretical distribution is fully specified along with the parameters as the critical region of the Kolmogrov-Smirnov test may not remain valid if the parameters are estimated from the data.

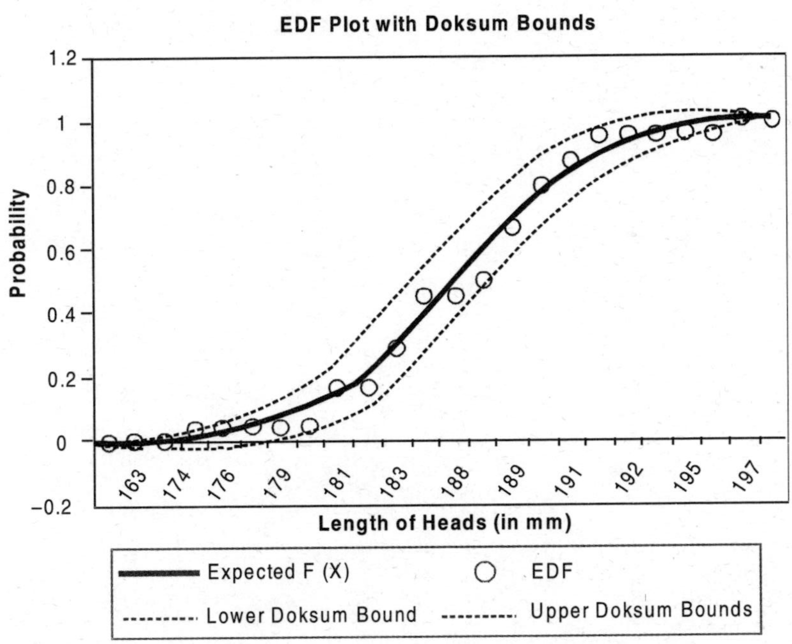

Figure 6.4: An EDF plot along with the Doksum Bounds

From the plot it appears that the data is a considerably good fit to the normal distribution.

REFERENCES

Anderson, T. W. (1958). *An Introduction to Multivariate Statistical Analysis*. New York: Wiley, 58.

Doksum, K. (1977). *Statistica Neerlandica*, 31, 53–68.

Fisher, N. I. (1983). Graphical methods in nonparametric Statistics: A review and annotated Bibliography, *International Statistical Review*, 51, 25–58.

Kotz, S. and Johnson, N. L. (1982). *Encyclopedia of Statistical Sciences.* Volume 1. John Wiley and Sons, Inc., 85–86.

Rice, J. A. (1995). *Mathematical Statistics and Data Analysis.* Duxbury Press, Belmont, California, University of California, Berkeley.

Chapter 7

Statistical Tests Using Graphical Tools

7.1 INTRODUCTION

The main purpose of Statistical Science is probably the analysis of data and to search for the facts hidden behind the figures. Statistical graphics being a part of the entire discipline must contribute sufficiently towards analysis of data in order to establish itself in the main stream of the subject. The graphical techniques are quiet often used for the analysis of data, especially for verification of assumptions on which statistical techniques are based. The use of graphics as analytic tools both alone or with other statistical techniques enhances the statistical sense of the user, and brings into notice certain abnormalities in the data set which would otherwise be concealed. In the words of J. W. Tukey- "All that seeing may be believing or disbelieving but above all data analysis involves visual as well as, statistical understanding."

The statistical summaries and test of significance are in a way data reducing techniques and by using them our attention get focused into the estimates or a single value calculated on the basis of a test of significance. Graphical displays can help us to visualize not only the central values and spread of the data but also supports comparison of all sorts for two or multiple data sets. Such techniques can be equally efficient, much simpler and appears with many added features compare to t-tests or ANOVA.

Most of our statistical analysis tools are based on several assumptions. Such tools can only be used in analysis provided the assumptions are strictly followed. In case of violation of assumptions we think of some robust statistical tools that may be of any help. But graphical techniques used for analysis is free from such assumptions and hence are most robust tools available. The concept of level of significance, confidence interval from parametric analysis can also be clubbed with the graphical techniques to make it more informative and reliable. In this chapter a discussion is made about several other plots that are used as diagnostic tools like comparing dispersions of dataset, randomness of data etc.

7.2 BIHISTOGRAM

7.2.1 Definition and Description

This is an important graphic tool, which mainly helps in the comparison of two data sets for location and dispersion. This can be used as a graphical alternative to t-test (for location) and F-test (for dispersion). In addition to this, the test can be used for the detection of outliers. The plot enables us to understand the distributional features like skewness and kurtosis of the two data sets that are plotted. This plot can be useful for assessing factors that has two levels. The plot can be of special importance in medical science to visualize the before versus after characteristics of a treatment. The two histograms when viewed separately gives the viewer an idea about probability models (normal, log-normal, etc.) for individual data sets, and to detect unexpected and unusual values in the data.

The plot consists of two histograms, one above the *X*-axis and one below the *X*-axis. The histogram above the horizontal axis is the histogram of the first variable and the one below the horizontal axis is the histogram of the second variable. A variation of the bihistogram is the relative bihistogram, where relative frequencies of both the data sets are plotted.

7.2.2 Working Data

The sample data for the plot is taken from Anderson (1958). The data gives the measurement (in mm) of head length of the first two adult sons in a few families.

Table 7.1: Head length of first and second sons of some families

Head length of first son(in mm)	191, 195, 181, 183, 176, 208, 189, 197, 188, 192, 179, 183, 174, 190, 188, 163, 195, 196.
Head length of second son(in mm):	179, 201, 185, 188, 171, 192, 190, 189, 197, 187, 186, 174, 185, 195, 187, 161, 173, 185.

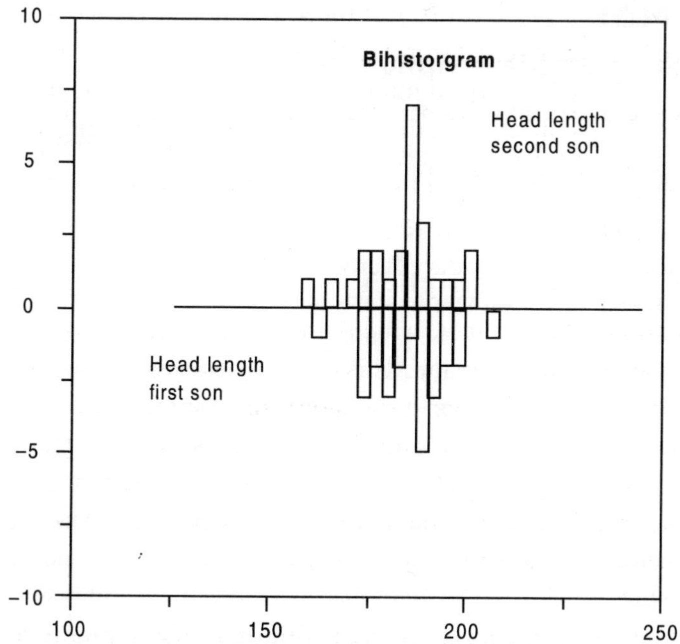

Figure 7.1: A bihistogram for the data in Anderson (1958)

Thus from the bihistogram we find that there is not much variability for the bars so far as their central values and dispersions are concerned. But peakedness appears to be more in case of Head length of second son then the first. It implies that the mode of the second series is more compared to the first. Both the series appears to be a little left skewed.

7.2.3 Axes

Above *X*-Axis: Histogram of the response variable for the first data series. For this data series we consider the head length of the first son.

Below *X*-Axis: Histogram of the response variable for the second data series. For this data series we consider the head length of the second son.

***Y*-Axis:** The frequencies are considered along this axis.

7.2.4 Uses

(a) The sample sizes do not need to be identical.

(b) The plot can be used for checking simultaneously several tests like- shift in location, shift in dispersion, change in symmetry/skewness, outliers, comparison of mode etc.

(c) As the plot is based on histograms, these are common and simple, these makes the plot easily understandable.

(d) The plot helps us to study the change in the probabilistic model (if any) of the two data sets in the before versus after cases.

7.3 BLOCK PLOT

7.3.1 Definition and Description

The block plot was initiated by Filliben and others (1993). This plot is generally useful if the primary factor has two levels only. The levels are coded with indices. This plot replaces the analysis of variance test by a less assumption dependent binomial test.

The *Y* axis of the plot consists of the response variable and the *X* axis consists of all combinations of the secondary factors. The graph comprises of some floating rectangles. The base of the rectangle starts at the lower value of the primary factor and ends at the higher value of the primary factor for a particular combination of the secondary factors. The codes for the various levels of the primary factors (which are generally integers) are written in the blocks in order to identify its position.

The figure below is a Block plot drawn based on the data given in Table 7.2. The grey vertical lines within the plot are used to partitions the different blocks. Each block consists of four rectangles one for each treatment. The values of the treatments at the two levels determine the position and length of the rectangles. The integers '1' and '2' within each rectangle is used to represent the relative position of the levels of the treatments.

Such a variety of comparison that is possible in a plot is otherwise absent in an ANOVA test. Instead of dealing with multiple numbers of groups if we deal with two groups only then the corresponding graphical technique can be treated as a graphical counterpart of t-test. One big advantage of drawing the graphs is that unlike the tests the graphical techniques are not based on any assumption.

7.3.2 Working Data

The working data is taken from Barlett (1936) which gives the values of number of surviving latherjackets for different controls and emulsions.

Table 7.2: Counts of surviving latherjackets for different Controls and Emulsions each at two levels

Block	Treatment	Emulsion 1	Emulsion 2	Emulsion 3	Emulsion 4
1	Level 1	6	17	8	12
	Level 2	10	8	11	17
2	Level 1	4	3	15	6
	Level 2	7	2	20	4
3	Level 1	4	6	10	12
	Level 2	12	3	7	10
4	Level 1	5	1	17	5
	Level 2	5	1	26	8
5	Level 1	2	2	14	12
	Level 2	6	5	11	12
6	Level 1	17	6	22	16
	Level 2	11	5	30	4

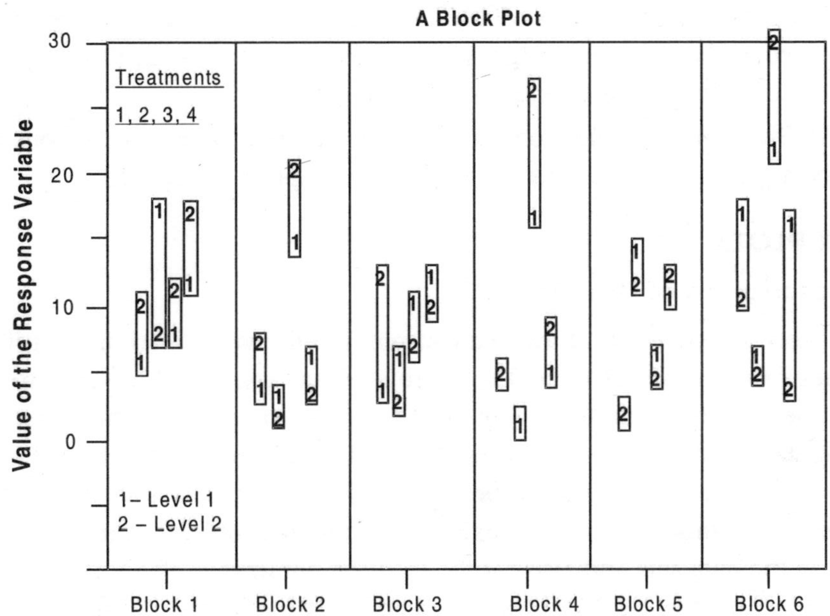

Figure 7.2: A Block Plot to the data in Table 7.2

From the diagram we can find that there are at least 11 cases out of 24 cases where the performance of Level 1 is better than Label 2. Also from the diagram we can find out that the performance of the various treatments has more variation in the fourth and sixth block. Also the variability is minimum in case of block 3. The third treatment under block 2, 4 and 6 appears to produce high response and so the third treatment can be considered as outliers in some of the blocks.

7.3.3 Axes

***X*-Axis:** It can represent all possible combinations of all factors. For this data set we consider the different blocks along the *X*-axis.

***Y*-Axis:** The vertical axis is used for representing the values of the response variable. For this case we took the number of surviving latherjackets along the *Y*-axis.

7.3.4 Uses

(a) The main purpose of the block plot is used to compare different levels classified for different treatments and in the different blocks. Thus we can judge whether there is any significant difference in the response variable for a particular factor under interest called the primary factor and if there is any significant change in the response variable for all other secondary factors (like, treatments, blocks).

(b) Also these plots can be used for the detection of outliers which is otherwise not possible in ANOVA technique. Once again for a complicated design the graphical representation appears to be much easy for any type of user but the techniques of ANOVA start becoming complex along with the complexity of the design and may even become difficult for people having sufficient mathematical background.

(c) ANOVA can be applied only if the data satisfies certain assumptions but for a graphical representation no such assumption is required.

7.4 BOX PLOTS

7.4.1 Description of the Plot

The box plots are simple to draw and can provide the visualization of at least five values from the set of data. These are median, upper and lower extreme values, upper quartiles and the lower quartiles. The box plots can be of use for producing data summaries and can also be used for data analysis. Here several data sets are plotted and we get a box for each of these data sets. The boxes when viewed in unison acts as an excellent graphical tool for comparing the variation in the data sets and also for the location parameter. The Box plot was introduced by Spear in 1952 and was taken up lately by Tukey (1970, 1977) and was popularized by him.

In order to draw a basic box plot for the representation of a set of observations one may abide by the following steps:

(i) To compute the median, the lower quartile and the upper quartile of the data.

(ii) The values of the variate are taken along the vertical axis and the category or serial number of the data set along the horizontal axis.

(iii) Plot the median, upper quartile and lower quartile. Draw a box (rectangle) between the upper quartile and lower quartile. The box represents the inter quartile range, that is 50% of the data that lies at the center. A line segment sections the box into two halves represents the position of the 50th percentile i.e. the median.

(iv) A line is drawn from the lower quartile to the minimum point and another line is drawn from the upper quartile to the maximum point. These two lines are called as the whiskers of the box plot. The box plot is thus also called as the box and whisker plot. One may draw a single box plot for each data set or a multiple box plot for all the data sets taken together.

7.4.2 Working Data

The data provided in Table 7.3 is self explanatory and the plot in Figure 7.3 corresponds to the Box plot to the data with the years as the categories.

Table 7.3: Index of Industrial Production : Sector-wise (Base Year : 1993–94 = 100)

Industry	1997-98	1998-99	1999-2000	2000-01
Mining	126.4	125.4	126.7	131.4
Manufacturing	142.5	148.8	159.4	167.9
Food Products	133.8	134.7	140.3	154.5
Tobacco	158.1	178.5	192.1	200.4
Cotton textiles	125.6	115.9	123.7	127.3
Wool, Silk	172	176.8	197.8	209.3
Jute	114.3	106	105	105.8
Textile	158.7	153.1	156.1	162.4
Wood	128.5	121.0	101.4	104.3
Paper	146.4	169.8	180.5	164.0

Source: Economic Survey, 2000-01.

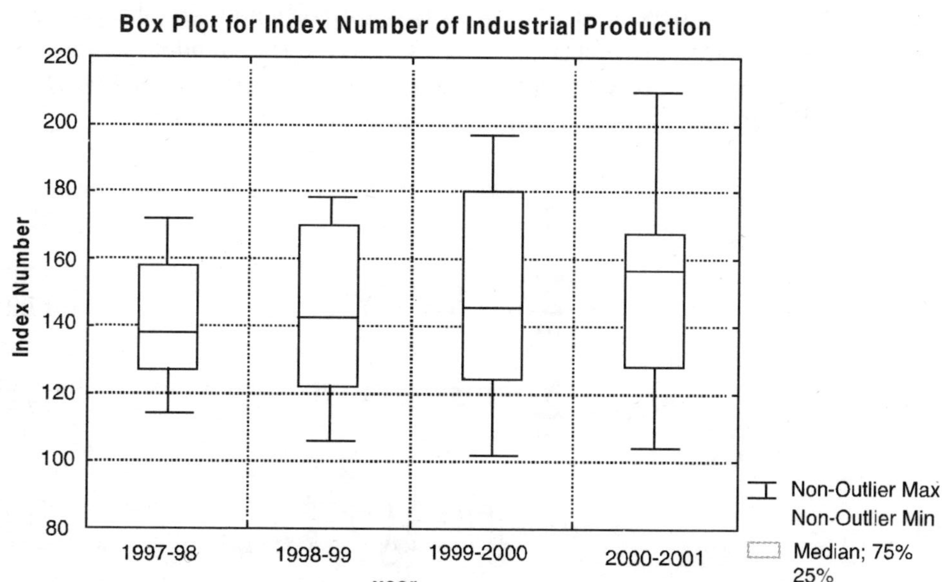

Figure 7.3: Box Plot for data provided in Table 7.3

From the plot we can say that the indices of 2000-01 has the maximum scattering, the same year also shows considerable shift in the median compared to the other years. However, both the location and dispersion appears to be minimum in case of indices corresponding to 1997 -98. Also, the indices are more asymmetric about the median in case of 1997-98 and 2000-01compared to the other two years.

7.4.3 Axes

X-Axis: It can represent the values of the variable that we suspect may have a relation to the response variable. For this data set we consider years along the X-axis.

Y-Axis: The vertical axis consists of that variable which we consider as the response variable. For this case we took the indices along the Y-axis.

7.4.4 Uses

(a) To compare the difference between the location (central value) of different series of observations.

(b) To compare the scattering of data in two or more sets of the observations.

(c) To understand the how much symmetrical a particular data set is about its median.

7.5 CHI PLOT

7.5.1 Definition and Description

Fisher and Switzer (1985) forwarded a plot termed as the chi-plot that can be used to detect the presence of dependence between two variables and also the pattern of dependence in the data if any. Let $(x_1, y_1), (x_2, y_2), \ldots, (x_n, y_n)$ be a random sample from H, the joint (continuous) distribution function for a pair of random variables (X, Y), and let $I(A)$ be the indicator function of the event A. For each data point (x_i, y_i), set

$$H_i = \sum_{j \neq i} I(x_j \le x_i, y_j \le y_i) / (n-1) \tag{7.1}$$

$$F_i = \sum_{j \neq i} I(x_j \le x_i) / (n-1) \tag{7.2}$$

$$G_i = \sum_{j \neq i} I(y_j \le y_i) / (n-1) \tag{7.3}$$

and

$$S_i = sign\left\{\left(F_i - \frac{1}{2}\right)\left(G_i - \frac{1}{2}\right)\right\} \tag{7.4}$$

On finding the values of $H_b F_b G_i$ we combine them to find the valuies of χ_i using the following transformation,

$$\chi_i = \frac{H_i - F_i G_i}{F_i(1-F_i)G_i(1-G_i)} \tag{7.5}$$

and

$$\lambda_i = 4S_i \max\left\{\left(F_i - \frac{1}{2}\right)^2, \left(G_i - \frac{1}{2}\right)^2\right\} \tag{7.6}$$

The the Chi-plot is drawn which is basically a scatteerplot between $(\lambda_b \chi i)$, provided $|\lambda i| < 4\left(\frac{1}{n-1} - \frac{1}{2}\right)$. The value of λi is a measure of the distance of the data point $(x_b y_i)$,from the center of the data set $(\tilde{x}, \tilde{y})$, where $\tilde{x}$ and $\tilde{y}$ are the medians values of $(x_b y_i)$, $i = 1, 2,..., n$ respectively.

It is always helful if the chi-plot is accompanied by a pair of horizontal grid lines at $c = -c_p / \sqrt{n}$ and here c_p is selected such that approximately $100p\%$ of the pairs $(l_b\ c_i)$ lie between the lines. The two lines acts as confidence bands to the chi-plot and can be used as a measure of independence. Fisher and Switzer (1985) found the c_p values for $p = 0.9$, 0.95 and 0.99 and the values are, 1.54, 1.78 and 2.18 respectively. Other values of c_p for different levels of significance can be derived using the Monte Carlo techniques. From the discussion it can be understood that the chi-plot depends only on the data through their ranks, where the plotted points are such that they should be approximately horizontal under independence.

7.5.2 Working Data

The data used for drawing this plot is a hypothetical one showing the values of two variates say *X* and *Y*. There are 28 pairs of observations. They are provided in the table below.

Table 7.4: Some hypothetical values of X and Y

X	32	56	67	3	43	21	43.9	89.12	75	57	34	22	54	52
Y	12	32	45	5	7	67	78	86	54	55	43	32	43	21
X	31	24	44	41	62	32	37	58	89	43	56	76	79	88
Y	43	56	79	21	43	23	78	98	75	65	54	32	33	33

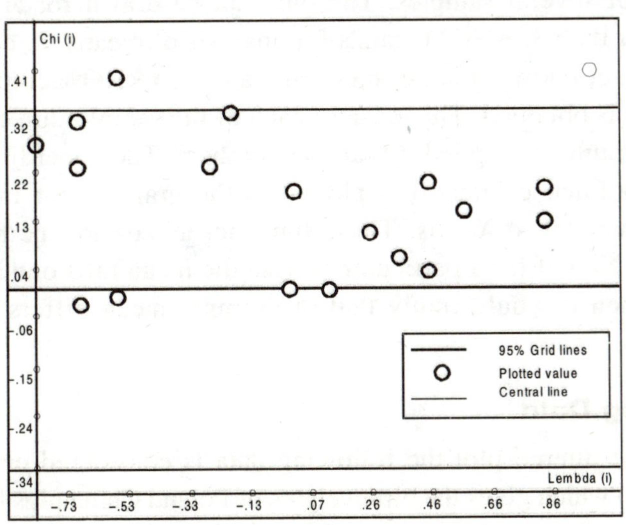

Figure 7.4: A chi plot for the data in Table 7.4

From the plot it can be inferred that the variables are dependent on each other as some points of the chi plot falls beyond the 95% grid lines. The decision can be taken with 95% confidence.

7.5.3 Axes

X-Axis: It can represent the values of the $\lambda_i = 4S_i \max \left\{ \left(F_i - \frac{1}{2} \right)^2, \left(G_i - \frac{1}{2} \right)^2 \right\}$,

where F_i, G_i are obtained using the equation (7.1), (7.2) and (7.3) etc.

Y-Axis: The vertical axis consists of the values of $\chi_i = \dfrac{H_i - F_i G_i}{F_i(1 - F_i)G_i(1 - G_i)}$.

7.5.4 Uses

(a) Provides a visual test for the dependence between two variables.

(b) Provides confidence intervals for such dependence which enables the viewer to differentiate if the dependence is significant for a given level.

(c) Can be used to study the randomness in a data set.

(d) Can be used to check if the residuals from a fitted model are randomly distributed and accordingly reach conclusions related to goodness of fit.

7.6 ANOM PLOT

7.6.1 Definition and Description

The chart is similar to the control charts. It is used for the multiple comparisons of the means of several samples. The plot can be drawn for any number of samples greater than 3. ANOM stands for analysis of means. In case the population mean is not known it is estimated and a $(1 - \alpha) \times 100\%$ confidence interval of the same is obtained. The mean of each of the samples are computed and plotted in the graph corresponding sample numbers. The upper limit and lower limit of the confidence interval is plotted in the graph paper in the form of straight lines parallel to X-axis. These lines act as confidence bands for the sample means. Now if for a particular sample the mean falls outside the confidence bands then it would imply that the sample mean differs significantly from the others.

7.6.2 Working Data

For drawing the jittered plot the following data is considered originally from Snedecor (1956) which gives the birthweights of Poland China pigs for four litters in pounds.

Table 7.5: Birth weight of Poland China Pigs in Pounds

	Liter 1	Liter 2	Liter 3	Liter 4
	2.0	3.5	3.3	3.2
	2.8	2.8	3.6	3.3
	3.3	3.2	2.6	3.2
	3.2	3.5	3.1	2.9
	4.4	2.3	3.2	2.0
	3.6	2.4	3.3	2.0
	1.9	2.0	2.9	2.1
	3.3	1.6	3.4	
	2.8		3.2	
	1.1		3.2	
Means	2.84	2.66	3.18	2.67

Overall Mean = 2.8628

Overall Standard Deviation = 0.6894

95% Confidence Interval for mean = [1.51, 4.21]

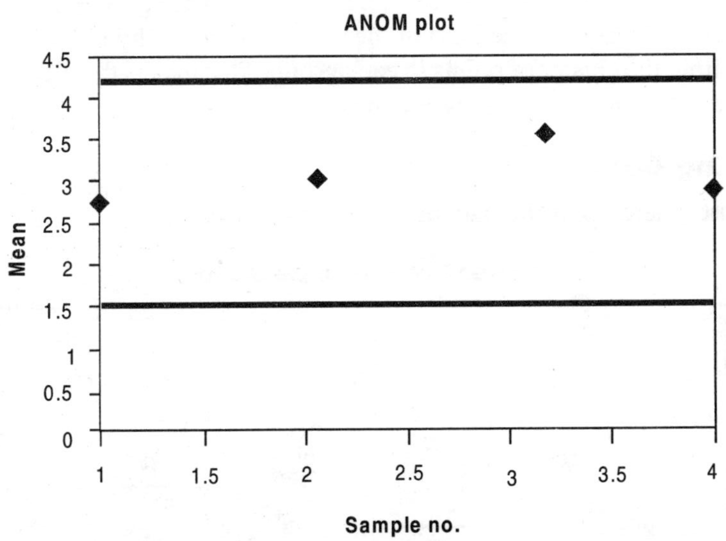

Figure 7.5: ANOM plot based on data in Table 7.5

From the plot we find that the sample means does not differ significantly at 5% level of significance.

7.6.3 Axes

X-Axis: The axis is used to represent the sample numbers.

Y-Axis: This axis is used to represent the values of the sample mean.

7.6.4 Uses

(a) The plot is simple to draw and easy to interpret.
(b) The plot is used for multiple comparisons of means from different samples.
(c) Even for unequal number of sample sizes this plot can also be used for comparison.
(d) The plot can be used as a quick check about the results of ANOVA for one-way classified data.

7.7 JITTERED PLOT

7.7.1 Definition and Description

A jittered plot is commonly used to represent multiple numbers of groups in a graph. The groups are taken along the *X* axis which are been separated by a distance. The groups may be named or some identifying number can be used. Above each of the group identifier variables (along *Y* axis) the values of the response variable are plotted in the form of small circles. The circles are kept transparent in order to avoid overlapping. Another simple technique of avoiding overlapping is to generate a random number between [– 0.2, 0.2] and add it to the group identifier variable before plotting. This disturbs the horizontal alignment of the points belonging to a particular group and hence acts a solution to the problem of overlapping in a group. The mean (or median) of the series is denoted by a small but thick straight line within the cluster of points in each group. This makes the means readily visible and the means across the groups can be compared.

7.7.2 Working Data

For drawing the jittered plot the data in Table 7.5 is considered.

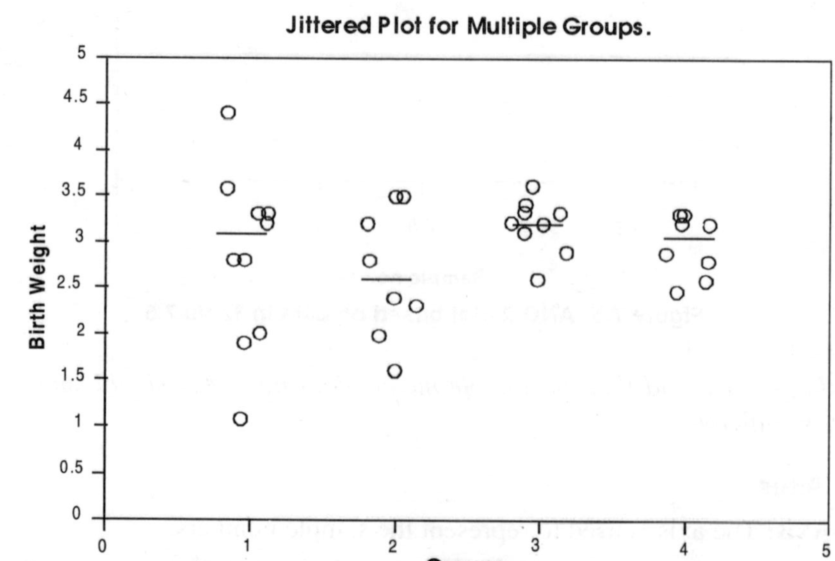

Figure 7.6: A jittered plot for data in Table 7.5

From the jittered plot drawn above one can understand that the means of the second group is less, considered to that of the other groups. Also the spread of the data is seen to be maximum in the first group and minimum in the last one.

7.7.3 Axes

***X*-Axis:** It represents the different groups. The groups are generally coded using some integers called the group identifier variables.

***Y*-Axis:** The vertical axis consists of the values of the response variable. Here the Birth Weights are taken along the vertical axis.

7.7.4 Uses

(a) The Plot is used for comparison of the central values of a number of groups.

(b) The Plot can also be used to view the spread of the data around the central value, range, symmetry around the central point and so on for each of the groups. Hence, the plot can be used as a tool for comparison of central values, deviation, range, concentration around central value and so on for multiple data sets.

(c) The plot can be drawn without the use of color.

7.8 RESIDUAL PLOT

7.8.1 Definition and Description

Whenever we express a variable with that of the other in terms of a mathematical relation then the most general form in Statistics is given by

$$y = f(x) + e$$

Where, the variable 'e' is called as the error component associated with the model. Gerenally, we will not know the values of 'e' for all the given values of 'x' Thus, we can estimate the coefficients involved in the function $f(x)$ and accordingly estimate the values of 'y' for a given 'x', which is denoted by '$\hat{y}$'. The difference 'y' and '$\hat{y}$' i.e. '$y - \hat{y}$' is called as the residual. In case of a good fit these residuals are randomly distributed. The residual plot is a plot between the independent variable i.e. x and the corresponding residual *i.e.* e.

7.8.2 Working Data

The data for this purpose is some hypothetical data for two variables X and Y.

Table 7.5: A hypothetical data for X, Y along with the trend values and residuals

X	*Y*	$\hat{Y} = 3.936X + 16.472$	*Residuals(e)*
1	13.06	20.4081	–7.3481

Contd...

2	18.44	24.3442	–5.9042
4	28.96	32.2164	–3.2564
6	39.16	40.0886	–0.9286
8	49.04	47.9608	1.0792
9	53.86	51.8969	1.9631
12	67.84	63.7052	4.1348
14	76.76	71.5774	5.1826
16	85.36	79.4496	5.9104
23	112.94	107.0023	5.9377
26	123.56	118.8106	4.7494
28	130.24	126.6828	3.5572
34	148.36	150.2994	– 1.9394
36	153.76	158.1716	– 4.4116
39	161.26	169.9799	– 8.7199

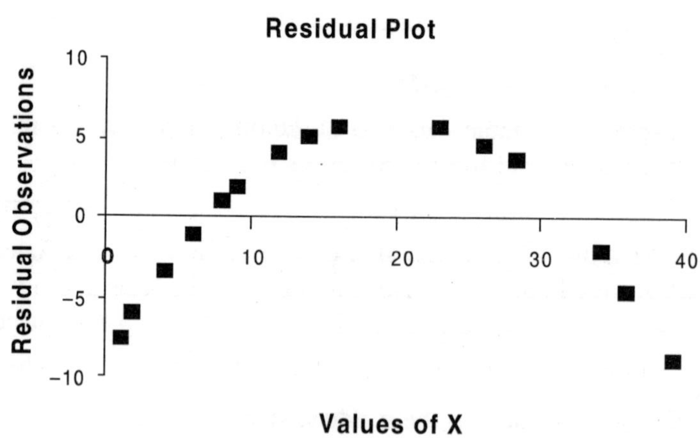

Figure 7.7: A Residual Plot for data in Table 7.6

The residual plot shows that there is a pattern in the residuals and hence it cannot be considered as random observations. This gives us an indication the model fitted is not a proper one for the data.

7.8.3 Axes

X **Axis:** The axis is used to represent the values of the independent variable.

Y **Axis:** This axis is used to represent the residuals from the fitted models.

7.8.4 Uses

(a) The plot provides a quick check for the randomness of the residuals from a model.

(b) The randomness of the residuals can be studied both for large as well as small samples.

(c) In case of non-randomness in the residuals the plot suggests a better model for the data.

7.9 RUN SEQUENCE PLOT

7.9.1 Definition and Description

A very simple graphical tool which produces a summary of a univariate data set related to location, spread as well as the presence of outliers is the run sequence plot. Let Y be a variable taking the values $Y_1, Y_2, \ldots, Y_n$. Then the plot is produced by plotting the values of Y_i against the index variable i. The points are plotted and are not joined. Thus we get n points of the data set separated by an equal horizontal alignment.

7.9.2 Working Data

The data is taken from Jarrett (1979) which gives the time intervals between successive coal mining disasters killing atleast 10 people from 15th March 1851 to 22nd March 1962. The actual data consists of 191 values but the table shows the first thirty values only.

Table 7.7: Data related to Coal Mining Disasters

157	123	2	124	12	4	10	216	80	12
33	66	232	826	40	12	29	190	97	65
186	23	92	197	431	16	154	95	25	19

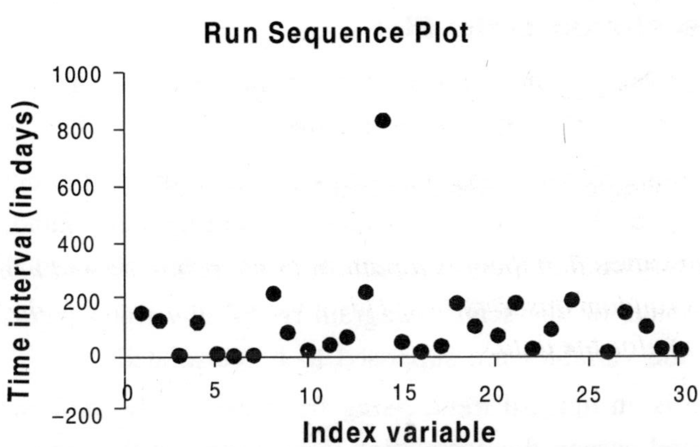

Figure 7.8: A Run Sequence plot for data in Table 7.7

The plot shows that the disasters are frequent with Y_{14} being an outlier. The location value remains almost uniform throughout the plotted values.

7.9.3 Axes

***X* Axis:** The index variables are taken along the X axis i.e. $i = 1, 2, 3, \ldots$

***Y* Axis:** The values of the response variable are considered along the axis. Here time interval (in days) are considered along the Y axis.

7.9.4 Uses

(a) The plot helps us to understand if there is any shift in the location of the data.

(b) The plot also helps us to understand if there is any shift in the spread of the data.

(c) The presence of outliers can also be detected.

7.10 SCATTER DIAGRAM

7.10.1 Description of the Plot

The scatter diagram is the simplest method of studying relationship between two variables, X and Y (say). The simplest device for ascertaining whether variables are related is to prepare a dot chart, where the horizontal axis represents one variable (X), and vertical axis representing the other (Y). Here we have the observations in pairs i.e. of the form $(x_1, y_1), (x_2, y_2), \ldots, (x_n, y_n)$; where, in each pair, the first value corresponds to X series and the other correspond to Y series. Each pair is then plotted in the graph in the form of a dot. The diagram so obtained is known as scatter diagram or dot diagram. From the scatter diagram one can have a fairly good idea about the relationship between variables. Scatter diagrams are two-dimensional diagrams. They are simple to draw and easy to interpret.

7.10.2 Interpretation of the Plot

The following points may be borne in mind in interpretation of a scatter diagram regarding the correlation between two variables:

(i) If the points are very close to each other, a fair amount of correlation may be expected between the two variables. On the other hand, if points are widely scattered, a poor correlation is expected between them.

(ii) If the points on the scatter diagram reveal any trend (upward or downward), the variables are supposed to be correlated.

(iii) If there is an upward trend rising from lower left hand corner to upper right hand corner, the correlation is positive. On the other hand, if there is a downward trend from upper left hand corner to lower right hand corner, the correlation is negative.

(iv) If all the points lie on a straight line starting from left bottom and going upwards to the right top, then the correlation is perfectly positive. On the other hand, if all the points lie on a straight line starting from left top and going downwards to the right bottom, the correlation is perfectly negative.

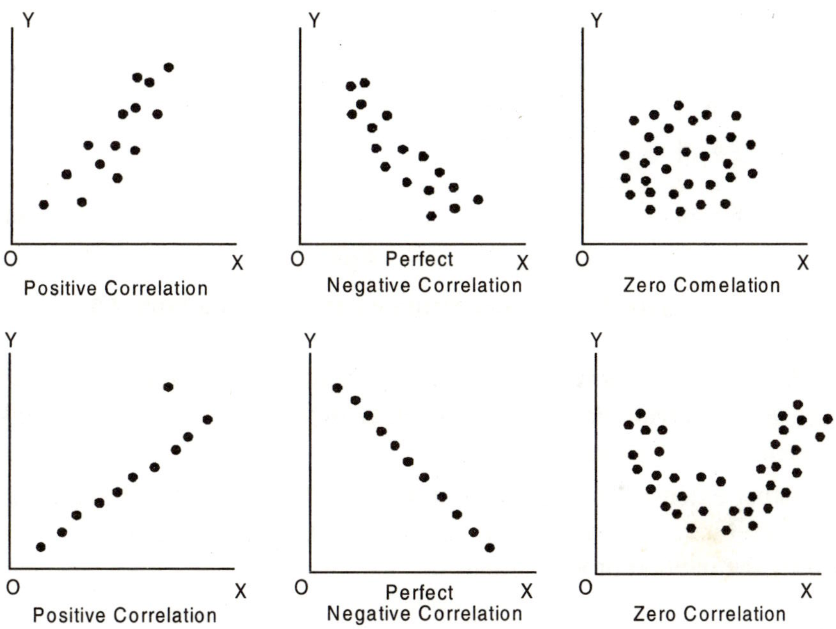

Figure 7.9: The scatter diagram for various types of correlation

7.10.3 Working Data

The working data is taken from Altman (1991). The data set gives the values of a fat content (in percentage) in the body of 18 individuals along with their ages.

Table 7.8: Percentage of fat in human body and corresponding age

Age	23	23	27	27	39	41	45	49	50
% of fat	9.5	27.9	7.8	17.8	31.4	25.9	27.4	25.2	31.1
Age	53	53	54	56	57	58	58	60	61
% of fat	34.7	42	29.1	32.5	30.3	33	33.8	41.1	34.5

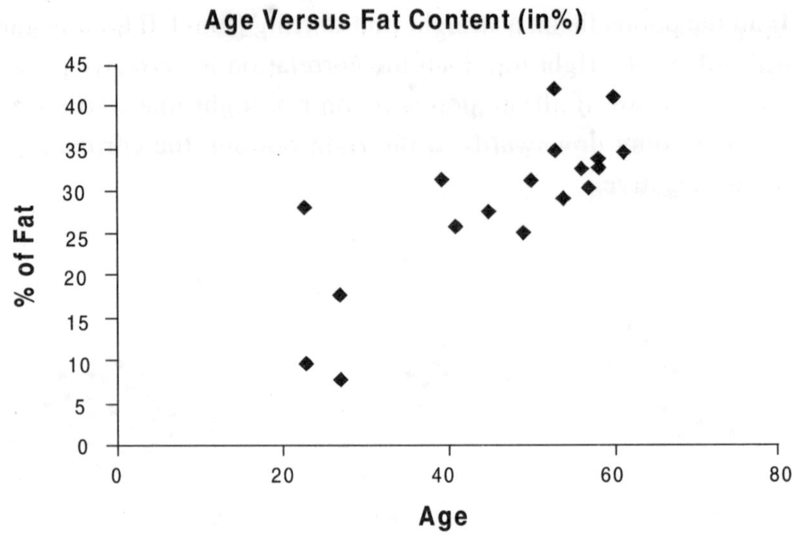

Figure 7.10: Scatter diagram based on data Table 7.8

From the scatter diagram we may conclude that there is a positive relation between the age of the individuals and the percentage of fat content i.e. with increase in age fat content is likely to increase.

Another extension to the scatter diagram is to plot the fitted line (or any other appropriate curve) along with the plotted points. The fitted line is generally obtained by the method of least squares.

7.10.4 Axes

***X*-Axis:** It represents the value of the independent variable. In the figure Age is independent and hence it is considered along the axis.

***Y*-Axis:** The vertical axis consists of that variable which we consider as the response variable or the dependent variable. For this case we took the fat content (in %) along the Y-axis.

7.10.5 Uses

(a) This is probably the most commonly used diagram to understand the relationship between two variables.

(b) The plot can be used as diagnostic tool for understanding the mathematical relation e.g. linear, quadratic, exponential, logarithmic etc. between the variables. Thus the user may fit the more appropriate curves instead of fitting all possible curves.

(c) Though not exact but this plot can be used for quickly deciding about the type of correlation between variables.

(d) Almost all the statistical software provides the option of drawing it.

7.11 CONCLUSION

This is definitely not the end but just a beginning. The graphical techniques should either stand alone or appear along with the commonly used techniques of data analysis. It may be accepted that the mathematical techniques are more efficient compared to the graphical techniques but if the graphical techniques can be plotted along with the significance levels as in case of the ANOM plot than the graphical techniques are much better. This is because the graphical tools actual reveals each and every data point making the comparison more extensive.

REFERENCES

Anderson, T. W. (1958). *An Introduction to Multivariate Statistical Analysis*. New York: Wiley, 58.

Barlett , M.S. (1936). Some notes on insecticide tests in the laboratory and in the field. *Journal of the Royal Statistical Society, Supplement*, 3, 185–194.

Filliben, J. J. (1997). *Dataplot Reference Manual.* Statistical Engineering Division, Information Technology Laboratory, NIST.

Fisher, N.L, and Switzer, P. (1985). Chi-plots for Assessing Dependence. *Biometrika*, 72: 253–265.

Snedecor, G.W. (1956). *Statistical Methods,* 5th edition, Ames: Iowa State College Press.

Tukey, J. W. (1970). *Explanatory Data Analysis (Limited Preliminary Edition), Vol. 1,* Ch. 5, Reading, Mass: Addison – Wesley Publishing Co.

Tukey, J. W. (1977). *Explanatory Data Analysis* (First Edition), Vol. 1, Ch. 5, Reading, Mass: Addison-Wesley Publishing Co.

Chapter 8

Time Series Plot

8.1 INTRODUCTION

Time series refers to a set of data where each of the observations are directly related to time. Thus plotting of the values of a time dependent variable along the y axis and the independent variable i.e. time along the x axis produces a time series plot. In subsequent sections of this chapter we will come across some of the plots which are not, as simple as that, but are related to time series analysis. A little reading of time series analysis, from any standard text is suggested, for better understanding of the terminologies expressed in the plots. However, attempt has been taken to explain the technical terms wherever needed.

8.2 AUTO CORRELATION PLOT

8.2.1 Definition and Description

Autocorrelation plot is commonly-used for checking randomness in a set of observations. This plot is specially used to check the presence or absence of trend in case of time series data. This randomness is ascertained by computing autocorrelations for data values at varying time lags. If random, such autocorrelations should be near zero for any and all time-lag separations. If non-random, then one or more of the autocorrelations will be significantly non-zero.

Let Y_i ($i = 1, 2, \ldots, N$) be a time series data. The formula for auto correlation of lag h is given by,

$$R_h = \frac{\frac{1}{N}\sum_{i=1}^{N}(Y_i - \bar{Y})(Y_{i=h} - \bar{Y})}{\frac{1}{N}\sum_{i=1}^{N}(Y_i - \bar{Y})^2}$$

The values of auto correlation are calculated for various time lags i.e. $h = 1, 2, \ldots$

The values of the various autocorrelation are then plotted against the corresponding lags. In case the data is random then the values of the autocorrelation will lie near to zero otherwise at least one autocorrelation will be significantly non-zero.

8.2.2 Working Data

The working data for this purpose is taken from *Newsletter for Pattern Recognition,* 13, No. 3, October 1990. The data gives the number of papers presented in International Conference on Pattern Reorganization, for the years 1981 to 1990 by American authors.

8.2.3 Axes

***X*-Axis:** In this plot along the X-axis we have taken the lags, which are various time lags i.e. $h = 1, 2, \ldots$

***Y*-Axis:** The values of the various autocorrelation are considered along the axes. Since autocorrelation lies between -1 to +1. So the axis takes the same range as its upper and lower limits.

Table 8.1: Table showing papers presented by American authors in International Seminars on Pattern Recognition

Year	1981	1982	1983	1984	1985	1986	1987	1988	1989	1990
Papers	52	52	81	55	134	63	154	65	62	150

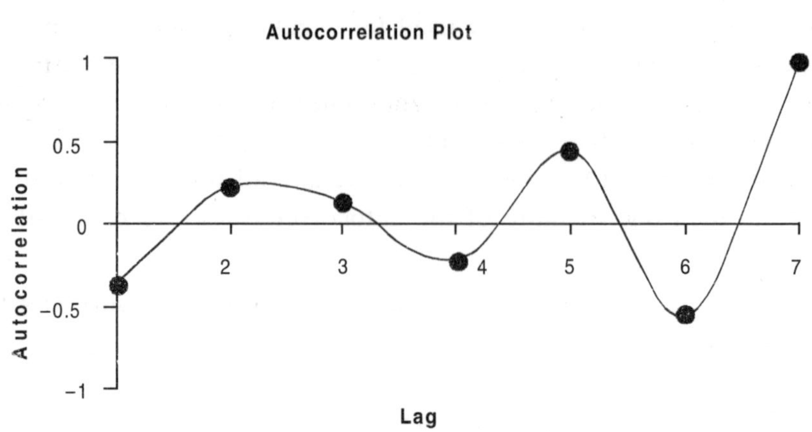

Figure 8.1: An Autocorrelation plot for the data in Table 8.1

8.2.4 Advantage

(a) The main advantage of the plot is to check if the data is random. Randomness is a gate keeper to many statistical tests and techniques. Thus the tool can be considered as a quick check to this feature.

(b) It provides a quick check whether the observations are related to an adjacent observation.

(c) The autoregressive nature in case of time series data can be put to check in this plot.

8.2.5 Disadvantage

(a) The plot calls for a lot of computation. The computations increases as the number of observations increases in the data set.

(b) It may become difficult to identify if the autocorrelation is significantly different from zero.

8.3 COMOVEMENT PLOT

8.3.1 Description and Definition

The plot can be used for the purpose of plotting the comovement coefficients far a particular time series for different lags or for two time series for different lags. The plot consists of the comovement coefficient along the Y axis and the lag along X axis.

The comovement coefficient is similar to the formula of correlation coefficient only instead of subtracting the mean from x_i and y_i, it is x_{i-1} and y_{i-1} that is been subtracted. The coefficient is given by,

$$C_1 = \frac{\sum_{i-2}^{n}(x_i - x_{i-1})(y_i - y_{i-1})}{\sqrt{\sum_{i-2}^{n}(x_i - x_{i-1})^2 \sum_{i=2}^{n}(y_i - y_{i-1})^2}} \text{ for lag 1.}$$

$$C_2 = \frac{\sum_{i-3}^{n}(x_i - x_{i-1})(y_i - y_{i-1})}{\sqrt{\sum_{i-3}^{n}(x_i - x_{i-1})^2 \sum_{i=3}^{n}(y_i - y_{i-1})^2}} \text{ for lag 2 and so on for lag } k.$$

The computed value of the coefficient lies between –1 and +1 and is useful in the study of comovement between two arbitrary time sequences. This procedure was originally proposed involving ARMA modeling of the two individual sequences.

8.3.2 Working Data

The table given below is taken from the Economic Survey, 2000-01 and shows the per capita availability of food grains net availability from 1990 to 2000.

Table 8.2: Net availability of cereals and pulses

Year	*Per capita availability per day (gms.)*	
	Cereals	*Pulses*
1990	435.3	41.1
1991	468.5	41.6
1992	434.5	34.3
1993	427.9	36.2
1994	434.5	37.2
1995	457.6	37.8
1996	443.4	32.8
1997	468.2	37.3
1998	417.3	33
1999	433.5	36.9
2000	426	32

Table 8.3: Comovement coefficients for different lags

Lag	1	2	3	4
Comovement Coefficients	0.7495	0.3214	0.6277	0.4632

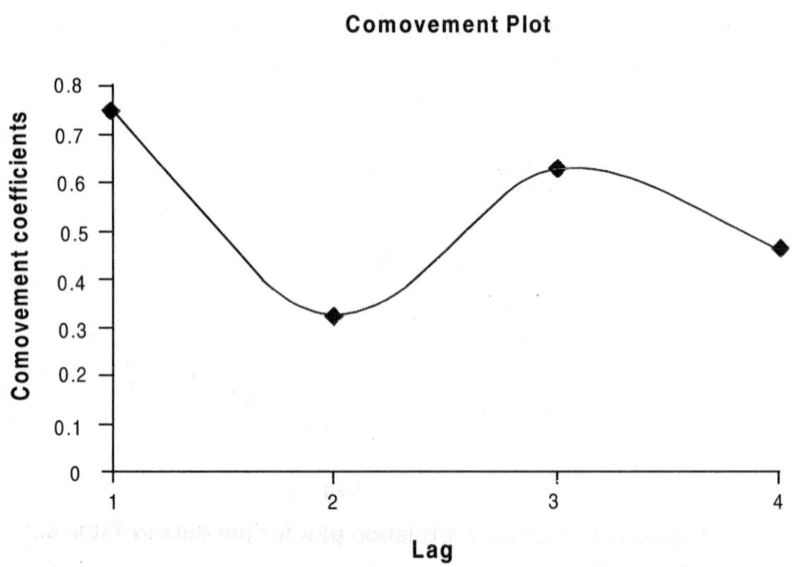

Figure 8.2: A Comovement Plot for the data in Table 8.2

8.3.3 Axes

***X*-Axis:** In the axis the lags are considered. So the value starts from 1 and may go at least up to k, where $k = n/4$, converted to nearest integer in case k is fractional.

***Y*-Axis:** In the vertical axis we consider the comovement coefficients for different lags. So the maximum range of the axes is from –1 to +1.

8.3.4 Uses

(a) To visualize the comovement coefficients for between two time series for different lags.

(b) To provide an idea about the type of model that is appropriate for the data.

8.4 CORRELATION PLOT OR CROSS-CORRELATION PLOT

8.4.1 Definition and Description

Cross-correlation is a measure of the degree of the linear relationship between two data sets. It is similar to autocorrelation except that it compares values in two different data sets instead of comparing different values within the same data set. Here the correlation between two time series for several lags is computed. They are then plotted in the graph. Thus the plot consists of the lags across the x-axis and the corresponding correlations along the Y-axis. The range of the Y-axis is from – 1 to +1. Thus a high correlation (either positive or negative) for a particular lag can be easily detected from the plot.

8.4.2 Working Data

The data used for this purpose is given in Table 8.2

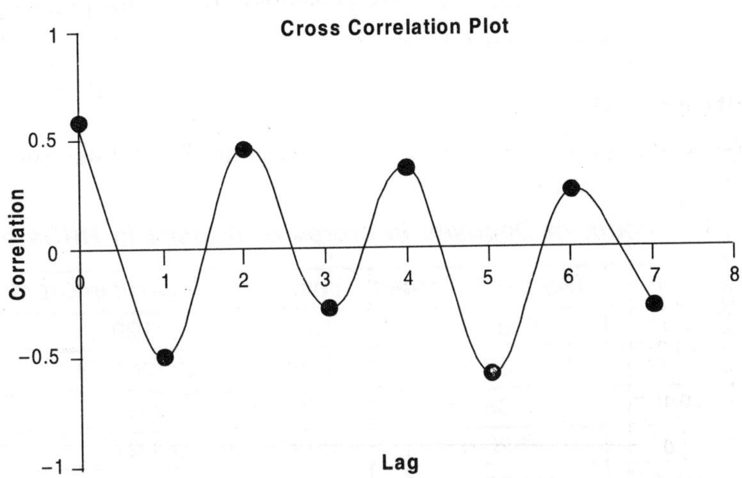

Figure 8.3: A Cross-correlation plot for the data in Table 8.2

The figure shows that at none of the lags the correlation is high. Even at lag 0 or at lag 1 where there is more chance of any relationship the correlation is not very high. Thus we may conclude from the plot that there is no linear relationship between availability of cereals and pulses for any of the lags.

8.4.3 Axes

X-Axis: In the axis the lags are considered. So the value starts from 1 and may go at least up to k, where $k = n/4$, converted to nearest integer in case k is fractional.

Y-Axis: In the vertical axis we consider the correlation coefficients for different lags. So the maximum range of the axes is from –1 to +1.

8.4.4 Uses

(a) The plot is used to study if two time series are independent of each other or not.

(b) The plot can be used to understand the dependence if a lagged series has some effect on another time series.

8.5 LAG PLOT

8.5.1 Definition and Description

A lag plot is more generally used in time series compared to other sub-field of Statistics. It is used to check whether a time series is random or not. Random data should not exhibit any identifiable structure in the lag plot. Non-random structure i.e. the existence of any trend would be understood by looking at the plot itself. Let Y_i $(i = 1, 2, \ldots, N)$ be a time series data. A lag plot of lag h is a scatter plot between Y_i versus Y_{i-h}. In case the values of the time series is dependent on the lag values, then the scatter diagram so plotted will show a clear pattern, otherwise one may conclude that Y_i and Y_{i-h} are independent of each other. In case the plot has lag 1 then the scatter plot will be between Y_i versus Y_{i-1}.

8.5.2 Working Data

The data relates to deposits (in crores of rupees) of all the regional rural banks of India taken together from 1991–2004.

Table 8.4: Amount of Deposits in crores of Rupees in RRBs of India

Year	*Deposit in Crores*	*Year*	*Deposit in Crores*
1991	4989.24	1998	22189
1992	5867.83	1999	27065.74
1993	6938.14	2000	30051
1994	8826.51	2001	37027
1995	11150.01	2002	43200
1996	14187.9	2003	50098.34
1997	18032.01	2004	56350.08

8.5.3 Axes

X-Axis: The values of the lag variable are taken along the axis. In this case since the lag is 1 so we have taken the values Y_{i-1}.

Y-Axis: The values of the time series are taken along the axes i.e. the values of Y_i.

8.5.4 Advantages

(a) The plot is simple to draw and easy to interpret.

(b) To check the randomness of the data.

(c) Existence of serial correlation in the data.

(d) Helps in deciding about a suitable model for a given time series data.

8.5.5 Disadvantages

Randomness is often a difficult task to perform by simple visualization. When the trend is very clear it is possible but this may not be the case always.

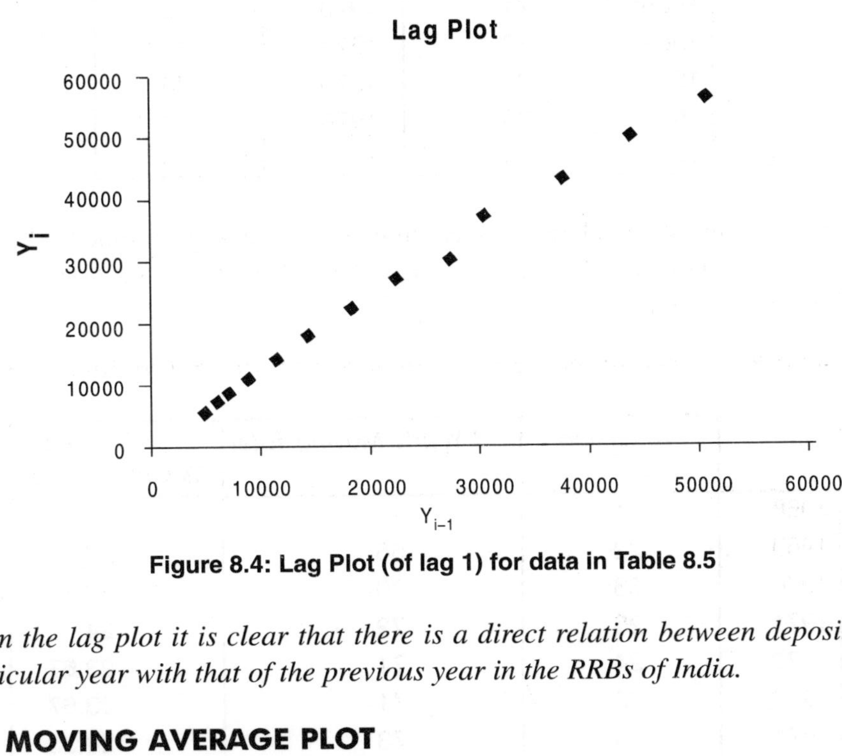

Figure 8.4: Lag Plot (of lag 1) for data in Table 8.5

From the lag plot it is clear that there is a direct relation between deposits of a particular year with that of the previous year in the RRBs of India.

8.6 MOVING AVERAGE PLOT

8.6.1 Definition and Description

Moving average method is a simple device of reducing fluctuations and obtaining trend values with a fair degree of accuracy. When a trend is to be determined by this method, the average values for a number of years (months or weeks) is secured. Then this average is taken as the normal or trend value for the unit of time falling at the middle of the period covered in the calculation of the period. The averaged values are then plotted corresponding to the different years. The points are then connected by a free hand smooth curve. Thus time is considered along the X axis and corresponding moving average values along the Y axis. Some times the original time series is also plotted in the form of a line diagram so that a comparison can be made with the graph of moving average.

8.6.2 Working Data

The data used for this plot is a hypothetical one representing data related to the production of an industry for different years.

Table 8.5: Year-wise production of an industry in thousand tons.

Year	*Product ion '000 tons*	*Year*	*Production '000 tons*
1968	21	1973	22
1969	22	1974	25
1970	23	1975	26
1971	25	1976	27
1972	24	1977	26

Correspondingly we calculate 3-year moving averages, determine the trend of the time series. For calculating the 3-year moving averages we construct the following table:

Table 8.6: 3-Yearly moving averages based on data in Table 8.5

Year	*Production*	*3 Yearly Moving Total*	*3 Yearly Moving Average*
1968	21	–	–
1969	22	66	22
1970	23	70	23.33
1971	25	72	24
1972	24	71	23.67
1973	22	71	23.67
1974	25	73	24.33
1975	26	78	26
1976	27	79	26.33
1977	26	–	–

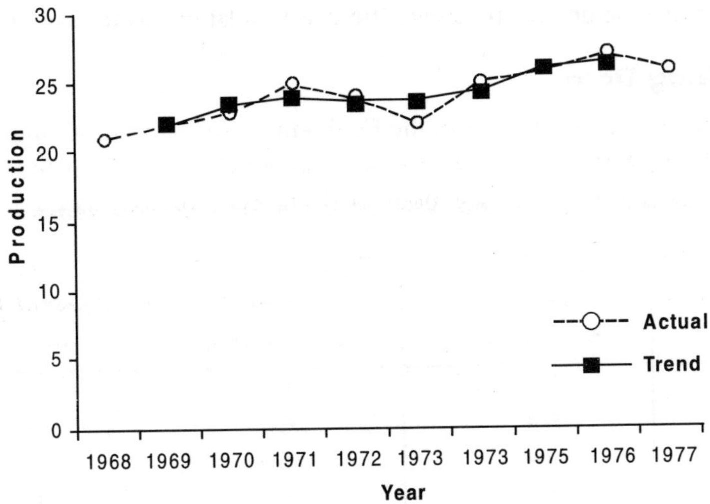

Figure 8.5: Moving Average Plot based on data in Table 8.7.

8.6.3 Axes

***X* Axis:** The axis is used to represent the time.

***Y* Axis:** This axis is used to represent the values of the response variable. The variable is Production in this case.

8.6.4 Uses

(a) The plot provides a visualization of the trend value of the data.

(b) A comparison between the actual values and the trend values may also be made using this plot.

8.7 MSE PLOT

8.7.1 Description and Definition

The MSE plot or the mean sum of square plot is a plot that can be used to locate the appropriate relation between a dependent variable and an independent variable. The plot can be used for time series data as well, to detect the most likely fitted model for a time series data from a host of given models. It can also be used to detect the same for a multiple number of time series based on the same time period.

Initially, we fit a linear trend between the response variable with the independent variable. Then the MSE (mean sum of squares) is computed for the trend by taking the differences between the fitted value and the actual data by squaring them and averaging over all the values of the independent variable. Proceeding in this way we calculate the MSE for other trend equations like, quadratic, exponential and so on. The graph is then drawn by plotting the MSE values (along the vertical axis) corresponding to the different types of regression (along the X axis). Thus the X axis consists of the names of different regression equations like linear, quadratic, exponential etc. and the Y axis consisting of MSE values.

The graph can be drawn for several time series data as well.

8.7.2 Working Data

The table given below is taken from the Economic Survey, 2001-02 and shows the gross national product, net national product and per Capita NNP of India.

Table 8.7: GNP and NNP of India for selected years

Year	GNP at factor cost (Rs. Crores)	NNP at factor cost (Rs. Crores)	Per Capita NNP (Rs.)
1951	5.7	5.4	3.6
1961	5.4	5	2.6
1971	6.4	6.1	3.6
1981	16.5	16.2	14
1982	11	10.4	7.9
1983	17	17.4	15
1984	11.9	11.6	9.2
1985	12.1	11.4	9
1986	11.4	11.1	8.8
1987	13.4	13.5	11.1
1988	19.3	19.6	17.1
1989	15.6	15.4	13
1990	16.5	16.7	14.3
1991	15	14.3	12
1992	14.3	14.1	12
1993	16.3	16.8	14.2
1994	17.5	17.5	15.2
1995	17.2	16.9	14.6
1996	16.1	16.1	13.9
1997	11.9	12	9.9
1998	15	15.5	13.3
1999	9.9	10.1	8.1
2000	7.9	7.8	5.9

Table 8.8: MSE for different types of regression based on data

	GNP	NNP	Per Capita NNP
Linear	11.3353	11.9716	11.3569
Quadratic	9.8508	10.6929	10.5625
Logarithmic	11.3097	11.9439	11.3569
Exponential	12.4115	13.1914	13.1044
Power curve	12.8164	13.69	13.7641

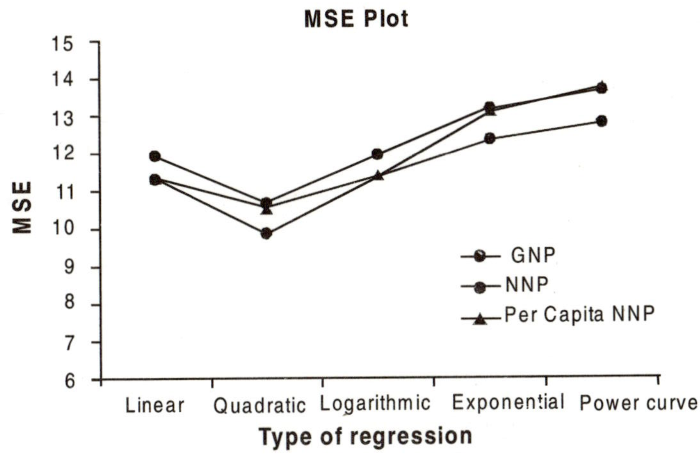

Figure 8.6: An MSE plot for the data in Table 8.8

From the MSE curves we find that for all the time series the quadratic equation is the best fit. Also the pattern of the MSE curves are same for all the time series.

8.7.3 Axes

***X*-Axis:** The axis consists of the names of different regression equations like linear, quadratic, exponential etc.

***Y*-Axis:** In the vertical axis we consider the corresponding MSE values of the trend equations are considered.

8.7.4 Advantages

(a) To visualize the MSE values for different types of regression equations of the same time series.

(b) To compare the distributional pattern of several time series data.

(c) Provides a tool for searching the best model for a given time series data.

8.7.5 Disadvantages

(a) Most of the commonly used statistical software does not provide the option to draw the plot.

(b) The calculations involved are very lengthy.

(c) The significant difference between the MSEs computed for different time series data cannot be computed.

Chapter 9

Three Dimensional Plots

9.1 INTRODUCTION

The most challenging part of graphics is in the display and visualization of the relationship among three or more variables. However there is always a tendency of representing three dimensional things in two dimensions. The artists from time immemorial have done it quiet successfully. The canvas or paper or wall, where the artists show their talent is actually two dimensional where the three dimensional figures that we see around us are been portrayed. The same principles can be adapted to display trivariate quantitative data (Jacoby, 1998). Multidimensional graphics starts with three variables and then goes to more than three variables. But, three variables have ample scope of being represented in two-dimensions as well. Thus, one can take up the issue, of the representation of three dimensional data separately, instead of mixing it with higher dimensional cases. Claveland (1993) separates the three dimensions and higher dimensions and termed the data situations as '*trivariate*' and '*hypervariate*' data, respectively. In this chapter we discuss some graphical techniques where we find how three dimensional data can be represented in two dimensions, and that is termed as dimension reduction.

9.2 THE BUBBLE PLOT

9.2.1 Definition and Description

The bubble plot is a graphical technique that can be used for representing three variables each of which takes numerical values. This plot can be considered as an extension of a scatter plot. Here we choose two primary variables that are represented along *X*-axis and *Y*-axis respectively. The (x, y) points are plotted in the graph paper but in the plotting symbols are circles. The radius of the circles is proportional to the value of a third variable. Precisely speaking the bubble plots represent the values of three variables by drawing circles of varying sizes at points that are plotted on the vertical and horizontal axes. Two of the variables determine the location of the data points, while the values of the third variable control the radius of the circles. The circles thus formed are kept colorless so that they remain transparent and the over lapping of points can be easily understood. Once the plot is drawn, the plotted points appear like bubbles of varying size. This explains the reason why the plot is named so.

9.2.2 Working Data

The data used for drawing the bubble plots is given in Table 9.1. The data comprises of heights, weights, age and sex of 30 individuals collected non-randomly from some known individuals.

Table 9.1. Data of 30 individuals pertaining to their height, weight, age and sex

Height (foot)	*Weight(Kg)*	*Age (Years)*	*Sex*
5.17	50	22	F
5.67	54	24	M
5	45	21	F
5.5	60	26	M
5.08	38	13	M
5.25	55	51	F
5.58	51	48	M
5.41	53	15	F
5.75	62	27	M
5.66	72.5	21	M
5.17	42	15	F
5.58	58	61	M
5.08	41	54	F
5	40	21	F
5.67	69	21	M
4.67	39	24	F
5.25	57	28	M
6	70	26	M
5.58	65	28	M
5.42	50	21	F
5.17	47	55	F
5.33	60	52	F
5.5	64	58	M
5.83	56	21	M

Contd....

4.75	50	12	M
5.25	72	40	F
5.5	73	31	M
5.58	71	61	M
5.08	54	59	F
5	57	28	F

The bubble plot in Figure 9.1 represents the first three variables of the table viz. Height, Weight and Age.

9.2.3 Axes

***X*-Axis:** It can represent the values of the variable that we suspect may have a relation to the response variable. For this data set we consider weight to be related to the height of the individual and hence we have taken height of the individual along the *X*-axis.

***Y*-Axis:** The vertical axis consists of that variable which we consider as the response variable. For this case we took weights along the *Y*-axis.

9.2.4 Purpose

(a) The graphs can be used to study the relationship between X and Y variables.
(b) It can also be used to study the variation between X and Y in presence of a third variable.
(c) It is helpful in detection of outliers.
(d) It can be used for visualization of three variables in two dimensions.

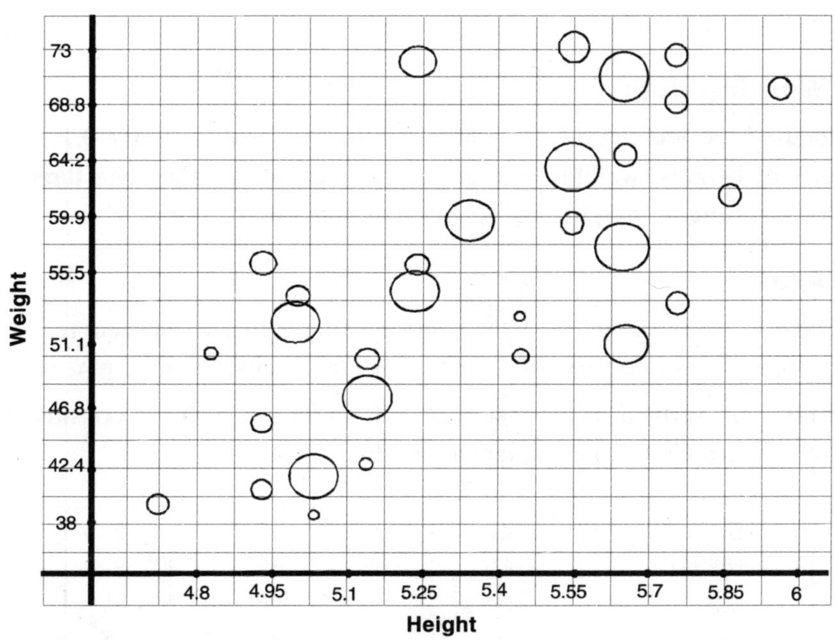

Figure 9.1. A bubble plot for the data in Table 9.1

9.2.5 Disadvantages

(a) This graphical technique requires some calculations for converting the value of the third variable to proportional radius.

(b) Bubble plot may sometime lead to over plotting. In case of many observations the bubbles may over lap either partially or entirely and hence make the representation untidy.

9.2.6 Future Direction

If color is used in the periphery of the bubbles then it can be used to represents another categorical variable. Here different colors are used for different categories and this can enable one to understand how any of the variables varies from category to category. For example, the display of sexes can be done for the Table in 9.1, by the use of color, say, red bubbles for male and blue bubbles for females.

9.3 CATEGORICAL SCATTER PLOT

9.3.1 Definition and Description

The categorical scatter plot is a graphical technique that can be used for representing three variables. Two of which are numerical values and the third one represents the categorical value. Here the two numerical values are represented along X-axis and Y-axis respectively. And the categorical values are given some codes. In this case females are represented by 1 and males are represented by 2.

9.3.2 Working data

The data used for drawing the categorical scatter plot given in Table 9.1. The data comprises of heights, weights, age and sex of 30 individuals. Here three of the variables viz. Height, Weight and Sex is represented in the graph.

9.3.3 Axes

***X*-axis:** It can represent the values of the variable that we suspect may have a relation with the response variable. So here we take height along the *X*-axis.

***Y*-axis:** The vertical axis consists of that variable which we consider as the response variable. So we take weight along the *Y*-axis.

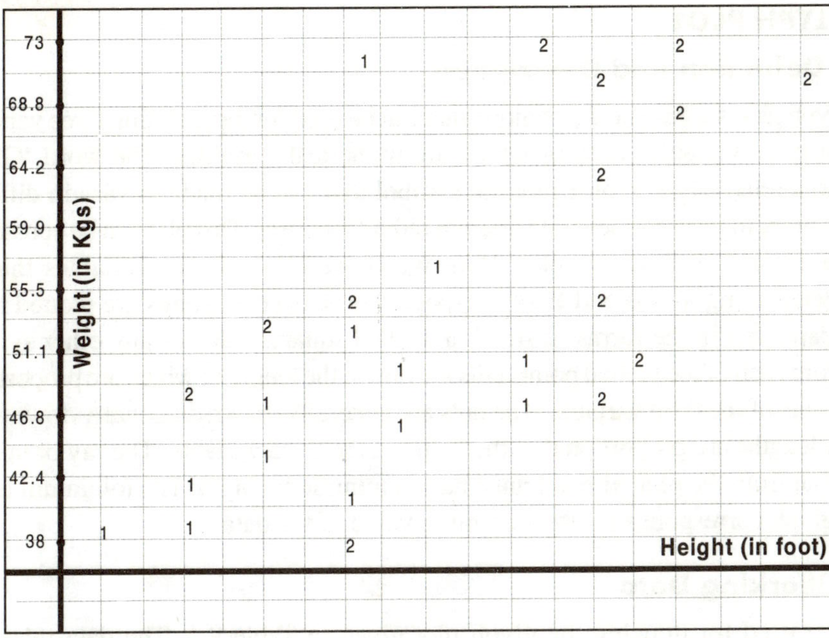

Figure 9.3: Categorical scatter plot for the data in Table 9.1

From the plot it is clear that there is wide variation in the height of the males and females as the '1' and '2' forms separate clusters with a different horizontal alignment. Also a difference between the vertical alignment between the clusters of '1' and '2' are also noticed. This implies that the two sexes also show difference in their average weight.

9.3.4 Purpose

(a) The graph can be used to study the relationship between X and Y variables.

(b) It can also be used to study the variation between X and Y in presence of a third variable. This plot enables us to visualize if there is any significant difference between the response variable or the independent variable across the various categories.

(c) It is helpful in detection of outliers.

(d) It can be used for visualization of 3 variables (two numerical variables and one categorical variable) in two dimensions.

9.3.5 Disadvantage

(a) In this plot only three variables can be represented, two numerical and one categorical. But any other variable be it numerical or categorical cannot be represented in this plot.

(b) With the increase in the number of categories under the categorical variable the plot becomes clumsy as different plotting symbols will be used making it difficult to interpret.

9.4 GLYPH PLOT

9.4.1 Definition and Description

The Glyph plot is a graphical technique that can be used for representing three variables or even more variables each of which are numerical variables. The word 'Glyph' refers to a relatively complex plotting symbol that can be made to vary in different ways in order to visualize several variables simultaneously. This plot can be considered as an extension of a scatter plot. Here we chose two primary variables that are represented along *X*-axis and *Y*-axis respectively. The (x, y) points are plotted in the graph paper but in the forms of small and filled squares (■) or any other symbol. Then from each of the plotted points tails are drawn the length of which are proportional to the value of the third variable. The tails are generally drawn diagonally downwards and the lengths are proportional to the value of the third variable. The ray or tail that comes out from the plotted point must have a minimum length. The minimum length of the glyph corresponds for the minimum value of the data.

9.4.2 Working Data

The data used for drawing the glyph plot given in Table 9.1. The data related to heights, weights, age of 30 individuals are considered. Here only three of the variables viz. Height, Weight and Age is used in drawing the graph in Figure 9.4. Thus the plot is used for representing three variables each of which is a numerical variable.

9.4.3 Axes

***X*-Axis:** It can represent the values of the variable that we suspect may have a relation to the response variable.

***Y*-Axis:** The vertical axis consists of that variable which we consider as the response variable.

The third variable is represented by a whisker that comes out from the plotted point and is proportional to its numerical value.

9.4.4 Purpose

(a) The graphs can be used to study the relationship between *X* and *Y* variables.
(b) It can also be used to study the variation between *X* and *Y* in presence of a third variable.
(c) It is helpful in detection of outliers.
(d) It can be used for visualization of 3 numerical variables in two dimensions and can be extended for representation of more variables.

9.4.5 Some Extensions to Glyph Plot

Though the basic glyph plot can be used to represent three numerical variables only it can be extended for the representation of more variables. The plotted points can be differently colored in order to use several categories of the data. Thus in order to attach the information related to sexes of the observations, one may use black color for representing the females and red color for the representation of the males.Thus the use of color in the glyphs plot, one can accommodate 4 variables (3-numerical and 1-categorical) and accordingly the category wise variation is obtained. This concept of use of color in glyph plot was advocated in Nicholson and Littlefield (1983).

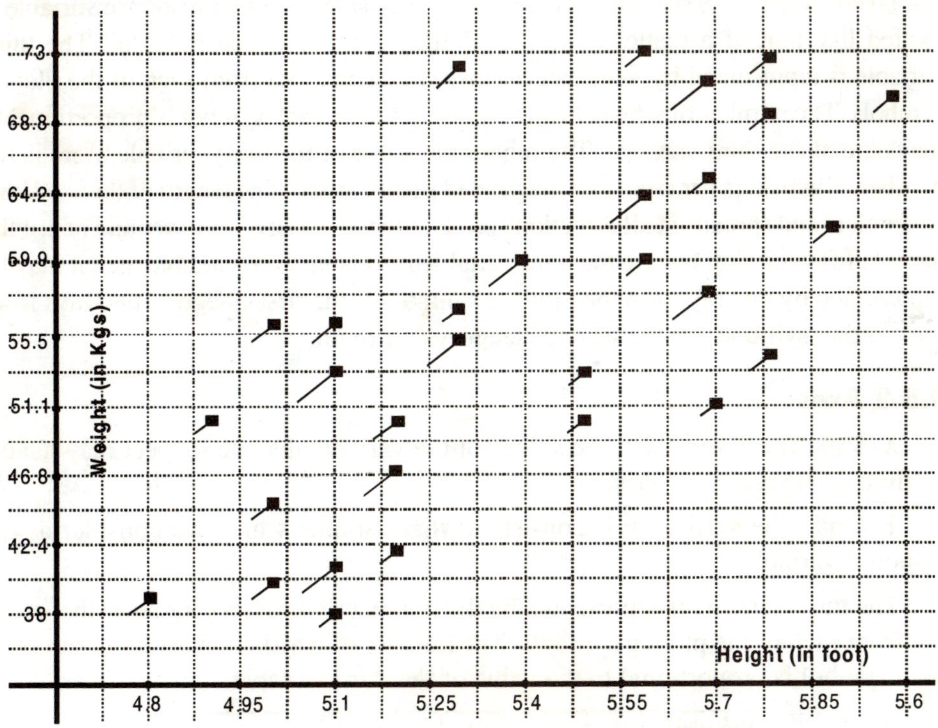

Figure 9.4: A glyph plot for the data in Table 9.1 representing Height, Weight and Age of the individuals

9.4.6 Disadvantages

(a) This graphical technique requires some calculation for converting the value of the third variable to proportional whisker from the plotted symbol in case one uses the length of the glyph for representation of the third variable.

(b) In case all the extensions are used in the glyph plot the plot will become very clumsy and in such a case the plot may become difficult to interpret.

(c) The length of the ray that comes out from the plotted point cannot be extended much as such extension may lead to over plotting. But as they are restricted in length so the variable represented along the ray (or tail) remains less informative and less quantitative.

9.5 SUNFLOWER PLOT

9.5.1 Definition and Description

The sunflower plot can be considered as an extension of a glyph plot. It was introduced by Cleveland and McGill (1984). The plot can be used to represent three variables two of which are numerical variable and the third one is either a numerical variable or a categorical variable. With the use of color one more categorical variable can also be displayed. At the very out set one independent variable and other response variable is plotted like that of a scatter plot. The plotting symbol is generally 'o'. The third variable is represented by whiskers (or rays) that come out from each of the plotted symbols. The number of whiskers that comes out from each symbol is proportional to the value of the third variable. The higher the value of the third variable more is the number of whiskers (or rays) that the plotted point carries. The length of the whiskers is identical and spreads in all directions producing the effect of a flower and hence the name sunflower plot. In case the third variable is a categorical variable then it may be represented by one ray from the plotting symbol for the first category, two rays from the plotting symbols for the second category and so on.

9.5.2 Axes

***X*-Axis:** It can represent the values of the variable that we suspect may have a relation to the response variable.

***Y*-Axis:** The vertical axis consists of that variable which we consider as the response variable.

The third variable is represented by the number of whiskers (or rays) that come out from each of the plotted symbols. The number of whiskers that comes out from each symbol is proportional to the value of the third variable.

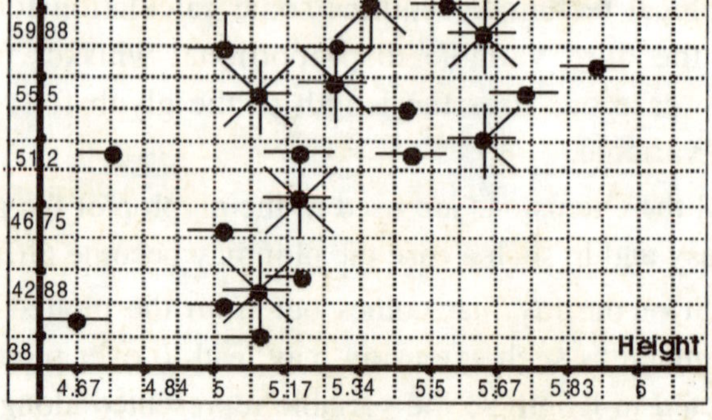

Figure 9.6. A sunflower plot for the data in Table 9.1

9.5.3 Advantages

(a) The graphs can be used to study the relationship between X and Y variables.

(b) It can also be used to study the variation between X and Y in presence of a third variable which may be categorical or numerical.

(c) It is helpful in detection of outliers.

(d) It can be used for visualization of 3 numerical variables or two numerical variables and one categorical variable in two dimensions. However with the use of color the plot can be extended for representation of more variables.

9.5.4 Disadvantages

(a) As the number of observations increases the sunflower plot may lead to over-plotting.

(b) Since the magnitude of the third variable is represented by the number of petals so for numerical variables the representation will not be much approximate.

9.6 CONCLUSION

In most of the graphical techniques discussed recently we find that the values of one of the variables are not scaled to be read from the graph and thus these graphs are less informative about that variable. The variables which are plotted along the X-axis/Y-axis are called as foreground variables and the remaining variables are called as background variables. In these plots the difference between observations in case of a particular background variable (for quantitative case) is often difficult for the eyes to determine precisely.

REFERENCES

Cleveland, W. S. (1993). *Visualizing Data.* Hobert Press, Summit New Jersey.

Cleveland, W.S. and McGill, R. (1984). Graphical perception: theory, experimentation and application to the development of graphical methods. *Journal of Americian Statistical Association*, 79, 531 – 554.

Jacoby, W. (1998). *Statistical Graphics for Visualizing Multivariate Data.* Sage University Papers Series on Quantitative Applications in the Social Sciences, 07 – 120. Thousand Oaks, CA: Sage.

Nicholson, W. L., and Littlefield, R. J. (1983). Interactive color graphics for multivariate data. In K. W. Heiner, R.S. Sacher and J. W. Wilkinson (Eds.), *Computer Science and Statistics: Proceedings of the 14th Symposium on the Interface.* New York: Springer-Verlag.